William Lant Carpenter

Energy in nature

With some additions

William Lant Carpenter

Energy in nature
With some additions

ISBN/EAN: 9783337259617

Printed in Europe, USA, Canada, Australia, Japan

Cover: Foto ©berggeist007 / pixelio.de

More available books at **www.hansebooks.com**

ENERGY IN NATURE,

BEING, WITH SOME ADDITIONS, THE SUBSTANCE OF A COURSE
OF SIX LECTURES UPON THE

FORCES OF NATURE AND THEIR MUTUAL RELATIONS,

Delivered under the auspices of the Gilchrist Educational Trust,
in the Autumn of 1881,

BY

WM. LANT CARPENTER, B.A., B.Sc.,

FELLOW OF THE CHEMICAL AND PHYSICAL SOCIETIES, AND OF THE
SOCIETY OF CHEMICAL INDUSTRY.

CASSELL & COMPANY, Limited:

LONDON, PARIS & NEW YORK.

[ALL RIGHTS RESERVED.]

1883.

To my Father,

WILLIAM B. CARPENTER, C.B., M.D., LL.D.,

F.R.S., F.G.S., F.L.S., &c. &c.,

WHO HAS DEVOTED MUCH OF THE ENERGY OF A LIFETIME

TO THE SCIENTIFIC EDUCATION OF THE PEOPLE.

I Dedicate this Little Book.

PREFACE.

THE object of the following pages is to give to those who have had little or no opportunity of studying the subject, some idea of the mutual relations existing between the various so-called "Forces of Nature," expressed in the simplest language, but at the same time, it is hoped, with sufficient scientific accuracy. Whenever it is impossible to avoid the use of technical terms, they will be carefully defined and explained, and the illustrations used will, as far as possible, be matters of common experience. An attempt is made in the earlier pages to lay down clearly the distinction between *Force* and *Energy*, and to explain how the language of the older books on the so-called Forces of Nature, the Correlation of Forces, &c., has been of late modified by the development of the doctrine of Energy and its Conservation. The book may be shortly described as an endeavour to expound in popular, yet accurate language, the meaning and consequences of that important principle known as the Conservation of Energy. Considerable pains, however, have been taken, especially in dealing with Electricity, to illustrate and explain the very latest developments of the subjects treated in the text, since the transformation of Mechanical into Electrical Energy by the dynamo-machine is a remarkably good example of the general principle. The substance of what is here written was delivered extempore by the writer to large audiences of artisans and others, under the auspices of the Gilchrist Educational Trust, in five Lancashire towns during the autumn of 1881, the lectures were abundantly illustrated by experiments, and by the projection of photographs upon the screen. It is believed that this was the first occasion on which the attempt was made to bring home to those who were not, in the ordinary sense of the term, either educated persons or students of science, the important practical consequences arising out of, as well as some of the glorious thoughts suggested by the consideration of, that grand doctrine of the Conservation of Energy, probably the most sublime generalisation of modern times, since its effects are not confined to our own planet, but pervade the Universe.

So much interest was excited on the subject among the audiences to whom the lectures were addressed, and so many enquiries have been made for a book in which the subject is treated in a popular manner, as to create a belief that the publication of the substance of the lectures in book-form will meet a real want; although no one, probably, is more conscious than the writer, of the different impressions produced by essays on a subject, and by the same matter treated orally, with experimental demonstrations and other illustrative aids to the comprehension of what may be to many readers comparatively unfamiliar scientific truths.

I wish to express my obligation to Professor Balfour Stewart, and to one or two other friends, for valuable suggestions made to me while the book was passing through the press; and to Messrs. Macmillan, Mr. Stanford, Messrs. Longman, and the publishers of *Engineering*, for permission to reproduce certain illustrations.

36, *Craven Park*,
    Harlesden, London, N.W.
   September 1st, 1883.

CONTENTS.

NOTE.—The following outline of the Course of Six Lectures, as arranged with the Gilchrist Trustees, will convey a fair idea of the contents of the six chapters, whose headings are given above. The delivery of the Course was prefaced with an Introductory Lecture by Dr. W. B. CARPENTER (the Secretary of the Trust), on "The Philosophy of a Lucifer Match," illustrating the connection between Mechanical Force, Heat, Light, Chemical Action, Electricity, and Vital Energy, all of them concerned in the simple act of "striking a light."

LECTURE I.

MATTER AND MOTION—FORCE AND ENERGY.

Different states of Matter (solid, liquid, and gaseous), all having weight, and offering Resistance—Indestructibility of Matter—Changes in its state.

Different kinds of Motion—Visible Motion of Masses—Invisible Motion of Particles, or Molecular Motion.

Different Modes of Force—Attraction of Gravitation giving Motion to Masses—Attraction of Cohesion holding together the particles of Masses.

Energy, or the Power of doing work—Energy of Motion, and Energy of Position, or Potential Energy—Relation of Energy to Mass and Motion—Its Disappearance really a change in its mode of manifestation.

and Electricity in motion —Transmission of Electricity by Conduction —Electric Currents—Heat and Light produced by opposing resistance to their passage—Mechanical effects of their interruption— Induced Currents — Ruhmkorff's Coil — Production of Chemical Action by Electricity—Law of Equivalence—Electrotyping and Electro-plating—Electrical Storage of Energy in Secondary Batteries.

LECTURE V.

MAGNETISM AND ELECTRICITY.

Magnetism originally derived from Lodestone; now induced in iron and steel bars by Electric Currents—Electro-Magnets—Mariner's Compass; its direction due to the Magnetism of the Earth— Deflection of Magnetic Needle by Electric Current—Induction of Electric Currents by giving Motion to Magnets—Magneto-Electricity.

Application of these facts in the Electric Telegraph, the Telephone, Microphone, and Photophone, and in the Dynamo-machine for Electric Lighting, and for the Transmission of Power.

LECTURE VI.

ENERGY IN ORGANIC NATURE.

Growth of Plants, and formation of their material, dependent on the Light and Heat of the Sun—Reproduction of this Light and Heat

ENERGY IN NATURE.

CHAPTER I.

MATTER AND MOTION; FORCE AND ENERGY.

BEFORE studying any of the so-called Forces of Nature, some of which are known under the names of Gravity, Chemical Attraction or Affinity, Electricity, Magnetism, Heat, and so on, it is desirable to say clearly and emphatically that these so-called Forces are not "things" in themselves, in the sense in which sand, wood, &c., are "things" in themselves, and that they are only known, and can only be investigated, by their effects upon substances, or, to use a more precise phrase, upon what the scientific man calls Matter. We cannot lay hold of some thing and handle it, or deal with it in any way, and say, "this is electricity," or "that is heat," in the sense in which we can say, "this is iron," or "that is clay." The so-called forces of nature have been well and truly spoken of as the moods, or affections, of matter. The relations of an individual to the objects which surround him, vary

B

with the mood in which he is; in a similar way,
the relations of any object in nature to other
objects in its immediate neighbourhood vary
with the mood of these objects, or with the way
in which they affect each other.   Thus, for
example, a leaden rifle-bullet weighing exactly
one ounce may be held in the hand and will
produce the sensation of weight and of metallic
coldness, but in other respects it appears to be in
a very impassive mood; if it be melted in an
iron spoon its weight will be unchanged, but
it can no longer be handled with impunity, and
is in a very heated mood; and again, if it be
fired off from a rifle it will still be the same
leaden bullet, but in a very destructive mood.
Lastly, if it be brought under the influence of a
current of electricity it will be found to behave towards iron and steel very much as a magnet does,
and it may be said to be in an attractive mood.

Hitherto we have spoken of Force, and of the
forces of nature; and so long as heat, chemical
attraction, electricity, magnetism, &c. &c., were
regarded as so many distinct and separate forces
(called in the older books on natural philosophy
the " imponderables," *i.e.*, the un-weigh-ables),
this phraseology served its purpose very well.
As, however, the idea gradually became developed
that these so-called various forces were all different manifestations of *a power of doing work* (*i.e.*,
causing change), residing in or acting through

matter, the need was felt for some general phrase which should include them all; accordingly the word *Energy* was adopted, as expressing more accurately this fundamental idea. Energy, then, is the *power of doing work*, in the strict scientific sense of the term. The phrase is an expressive one, and the consideration of what is implied in the phrase " an energetic man " will assist the reader in grasping the idea implied in it. We do not exactly know what energy is, but we recognise it, just as we recognise life (about the nature of which we are equally ignorant) in various forms, and, as will presently be seen, we can measure it very exactly. On this view then all the so-called forces of nature, or the various moods that affect matter, are so many kinds of energy, which is capable of assuming various forms, and of being changed from one form to another by apparatus arranged for the purpose by man, but is *never created afresh or destroyed entirely*, by any contrivance of his. This is the idea intended to be conveyed by the modern phrase, " The Conservation of Energy " (in place of that of the " Correlation of Forces"), which is the subject of the following pages, an idea which the diagram, Fig. 2 (p. 13) is designed also to convey to the eye.

Let us consider for a moment what is implied exactly by the term Force. A little thought given to the subject will show that whenever it is de-

B 2

sired to set anything in motion, or to bring to rest
anything already in motion, the use of force is
necessary. The exertion required to bowl, and
to catch, a ball at cricket, and the action of the
locomotive and the brake in the railway train,
are familiar examples of this. Hence, Force may
be defined as *any cause which alters or tends to
alter a body's natural state of rest, or of uniform
motion in a straight line*, and it should be borne
in mind that the states of rest and of motion here
referred to are not merely molar, but also molecular,
i.e., not merely motion of the body as a whole,
but of the motion among themselves of the very
minute particles or molecules (*vide* pp. 17–20) of
which the body is made up. There is very good
reason to believe that all the various energies of
nature, light, heat, electricity, magnetism, &c.,
are derived from different kinds of molecular
motion. Prof. Tyndall wrote a book several
years ago entitled " Heat considered as a Mode
of Motion." Prof. Hughes has quite lately (February, 1883) exhibited to the Royal Society some
very curious experiments, which seem to confirm
the idea long entertained by Sir W. Thomson
and others that magnetism is another form of
molecular motion. A wire along which an
electric current is made to pass undergoes no
change in weight thereby, nor, if it is large
enough (Chap. IV.) in any outward appearance,
but under certain conditions (in a dynamo-

machine, for example, Chap. V.), it can produce
not only molecular motion, but even that kind
of motion familiar to us as mechanical.

It is evident, therefore, that Force also can
only be known by its effect upon Matter, and that
it is not a " thing " (in the ordinary sense of that
word) any more than the bank-rate of interest is
a sum of money, or a birth-rate of children is
the actual group of children who are born in a
year. Force is simply the expression of the
rate or speed at which any change takes place
in matter; what its essence, or primordial cause
is, is a problem that science does not attempt to
solve.

The terms Energy and Force do not, therefore,
mean exactly the same thing, and, indeed, it is
most important to grasp, and to bear in mind,
the distinction between them. Energy involves
two distinct ideas combined, whereas Force involves only one. Energy has been defined as
" the power of doing work," and work is force
exerted through space, *i.e.*, the idea of motion
of some kind is connected with it.* The distinction again between energy and what the

* When a weight rests on the ground, the weight pushes the
ground down with a certain force, and the ground pushes the weight
up with the same force. It is obvious, however, that a weight in
this condition is incapable of doing any work, *i.e.*, of producing
motion of any kind; indeed, the very expression "dead-weight"
applied to it is a practical admission of its inefficiency in this respect. The pressure that this inert weight produces is force and
not energy, but the operation of raising the weight is doing work,

artizan and engineer recognise as "power" is essentially difficult, and lies probably in the idea of direction—any form of directed energy is power.   In some lectures on the Transmission of Energy, recently delivered before the Society of Arts, Prof. Osborne Reynolds compared the difference between undirected and directed energy, to the difference between a mob and a trained army, the individuals in whom the energy resided being, in both cases, the molecules (p. 20) or ultimate particles, of matter.

Reference has been frequently made to the term "Matter." What is this Matter, in the different moods or affections of which we recognise what is now known as Energy? It may be stated in general terms to be "whatever can affect one or more of our senses;" but there are certain kinds or conditions of it for which we need to increase the sensitiveness of what have been called our "five gateways of knowledge" by suitable apparatus, in order to become aware of this affection of our senses. Thus, for example, the sense of sight may be assisted by magnifying glasses, and those of touch and hearing by certain electrical contrivances, such as the telephone and the micro-

and involves the expenditure of energy. In every case in which force is said to act, what is really observed is a transference (or a tendency to transference) of energy from one portion of matter to another; and the so-called force in any direction is simply the rate of that transference.

phone (Chap. V.), with which it is possible to
hear the footsteps of a fly crawling upon a piece
of wood, and to converse in spoken language
with a friend a thousand miles off.*

A little attentive consideration will show that
we usually meet with matter in one of three
states, known as solid, liquid, and gaseous.† It
will presently be shown that, in the majority
of instances, the occurrence of any substance in
one form or the other depends upon the temperature and the pressure to which the matter
is subjected. A very familiar example of the
three states of matter is to be found in ice,
water, and steam, each of which forms of water
can be changed into either of the others by
regulating the amount of heat to which it is exposed, while it still remains through all these
changes the same compound of two measures
of hydrogen with one measure of oxygen (*vide*
page 68), which the chemist recognises as
water by its behaviour towards other kinds of
matter.

It will be desirable now briefly to consider
one or two properties which are possessed, without any exception, by matter in all its states.
A stone left to itself in the air falls until it
touches the ground; a round stone or ball rolls

---

* Vide *Nature* for March 22nd, 1883.
† For the purposes of the present enquiry the question of the
existence of a fourth state of matter need not be discussed.

down a sloping surface; a mass of water, such as a stream or river, flows from a higher to a lower level in the channel in which it is confined; clouds appear to rise in the air, and descend again as rain; all these occurrences are due to the fact that all matter possesses *weight*. That this is the case with solids and liquids is a matter of such common experience as to need no more than a reference to it; but it is not so apparent, although equally and universally true, in the case of gases. It may, however, be readily proved by very simple apparatus (a balloon for example), and may be shown more exactly by weighing different kinds of gas in a glass-bottle provided with a stopcock, attached to one end of a delicate balance or pair of scales, as in Fig. 1. If all the air be removed from the bottle by an airpump, and the bottle be then carefully weighed, it will be found that when the stopcock is opened, and air is admitted into the bottle, weights must be put in the other scale-pan to restore the balance.

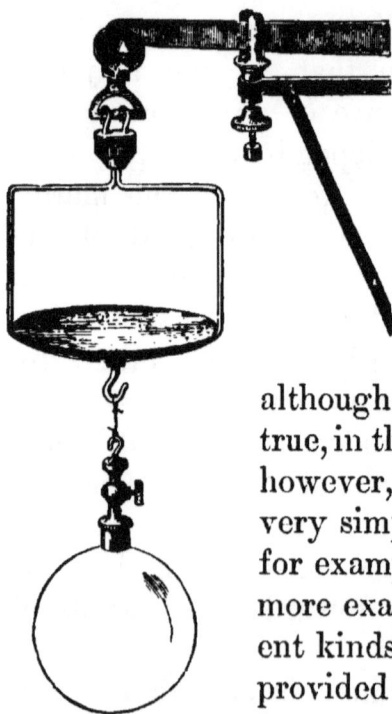

FIG. 1.

In this way it can be shown that five gallons of common air weigh very nearly an ounce, or, in the so-called metrical system in use on the continent, not only by scientific men but also in all the dealings of daily life, that one litre of air weighs 1·29 gramme. Instead of common air other gases may be allowed to enter the bottle, being conveyed there by a pipe, and it can thus be shown that they differ greatly in weight. Thus:—

100 cubic inches of air weigh	.	.	. 31 grains.
100 „ „ of carbonic acid weigh	.	47 „
100 „ „ of hydrogen weigh only	.	2 „

Now here we have an illustration of the fact—which is of considerable importance in connection with this property of weight, that substances differ largely in the mass or quantity of matter contained in the same measure thereof. Thus, for example, a gallon of water weighs exactly 10 lbs. (and in the metrical system a cubic centimetre of water weighs exactly one gramme), while a gallon of oil, or of spirit, weighs only 9 lbs., or less. If the weight of one-tenth of a gallon of water (or 27·27 cubic inches) which is 1 lb., be taken as a standard, and compared with the weight of the same measure of other substances, we shall find that the weights of equal measures, expressed in lbs., give us figures like the following :—

Hammered Copper	. 8·9	Brick . . .	. 2·1
Gold . . .	. 19·6	Clay . . .	. 1·9
Cast Iron .	. 7·0	Chalk . . .	. 2·3
Wrought Iron .	. 7·7	Dry Sand . .	. 1·4
Lead . . .	. 11·4	Glass . . .	. 2·7
Silver . . .	. 10·5	Spruce Fir . .	. 0·6
Platinum . .	. 21·5	Oak . . .	. 0·8
Tin . . .	. 7·4	Proof Spirit . .	. 0·9
Zinc . . .	. 7·0	Ether . . .	. 0·7
Mercury . .	. 13·6	Oil . . .	. 0·9

The annexed drawings of spheres (Fig. 2), and figures, also present the same relationship between other substances :—

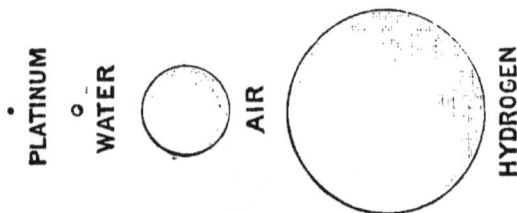

FIG. 2.

Hydrogen .	1		
Air . . .	14·4	1	
Water . .	11,943	829	1
Platinum .	256,764	17,831	21·5

Thus, for equal bulks, the metal platinum is 21½ times as heavy as water, 17,831 times as heavy as air, 256,764 times as heavy as hydrogen, and so on.

This relation between the measure (or volume) and the weight of any substance is known as its specific gravity, and is recognised in common life by such phrases as "Oil is lighter than water," "Lead is heavier than iron," &c., the words

" heavier" and "lighter" really implying the addition to them of the phrase " bulk for bulk."

Another characteristic of all kinds of matter is that it offers resistance. This, again, so far as our sense of touch is concerned, comes within our common experience, even in the case of gases, for who has not experienced the resistance offered to his progress by air in violent motion? Another mode of expressing the same thing is to say that matter possesses impenetrability, which means that no two portions of matter can occupy the same place at the same time. An amusing story illustrating this is told of King Charles II., in whose reign the Royal Society was founded. The king propounded to the philosophers composing it the following query:—"Why is it that when two gold-fish are put into a bowl of water full up to the brim, the water does not overflow?" The question was gravely debated at considerable length, until some one suggested that the question should be put to Nature, or in other words, that the experiment should be tried. The result, of course, showed that the " Merry Monarch " had been amusing himself at their expense.

It may be pointed out here that by no process at the command of man can matter be either created or destroyed. It is a most important truth that the utmost man can do in this direction is to change the combinations of matter. For example, when coal is burnt in the fire it is not destroyed, but in the act of burning (Chap. III.) its carbon and hydrogen unite with the oxygen of the air, forming gases which pass away into the atmosphere. The main object of this little book is to show that the same thing is true of the so-called Forces of Nature, that is to say, of the forms of Energy, which can neither be created nor destroyed by man, whose power over them is limited to changing their form, or mode of manifestation. This doctrine may be looked upon as the outcome of the scientific work done during the last thirty years by such men as Meyer, Helmholtz, Clausius, and others, in Germany, and as Grove, Joule, Thomson, Balfour Stewart, Tait, Tyndall, and W. B. Carpenter in England. Properly regarded in its various aspects, it is one of the most sublime generalisations of modern times ; the most stupendous phenomena of the universe are seen to happen in accordance with it, while the life and movements of the minutest living being, and the most insignificant natural phenomenon occurring in human experience, alike bear witness to its truth.

That when Energy seems to be destroyed, or to disappear, it does not really do so, but, like the magician in the "Arabian Nights," only vanishes to reappear under some other form (or mode of manifestation) it will be the object of the following pages to prove. A careful inspection of the

accompanying engraving (Fig. 3) may assist the
reader in grasping this grand idea, to which the
name "Correlation of the Physical Forces" was

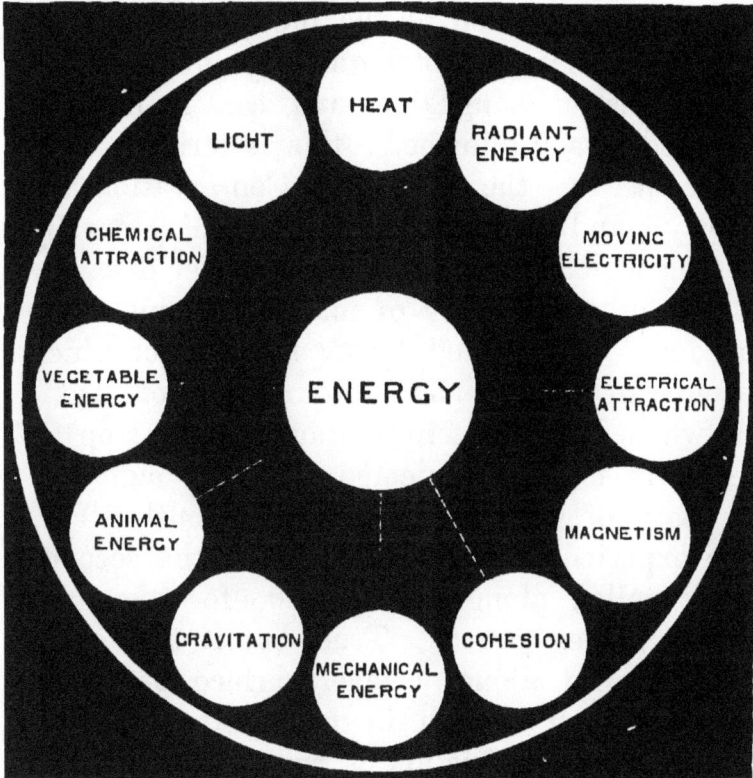

Fig. 3.

first given, but the progress of research has shown
(Chap. VI.) that it may be extended also to the
so-called Vital Forces, *i.e.*, to the Energy dis-
played in the phenomena of Vitality. The circles
are intended to represent movable discs, upon
which are written the names of some of the forms

of energy existing in Nature. It is obvious that any two of these discs may be made to interchange places, and thus to express graphically the change or transformation of one form of energy into another. A grand thought, which follows from the recognition of the indestructibility of energy, is that the total quantity of it in the universe is unchangeable, and can neither be increased nor diminished; this is strictly what is meant by the phrase " Conservation of Energy," and it is expressed in the diagram by the ring which surrounds all the smaller discs.

One other property of matter remains to be considered—that called *Inertia*, in virtue of which matter can neither start into motion of itself, nor, when it is once in motion, can it stop of itself. It may be objected that although the first part of that sentence is obviously true, daily experience tends to contradict the second. A ball rolled along the ground, for example, speedily comes to rest. True, but if that same ball be rolled along a level surface of ice it will travel very much farther, and the more the resistance of friction is removed from any body in motion, the longer will that motion continue. If a strong mental effort be made to put aside the impressions caused by the familiar facts of the case, it will then be evident to the accurate thinker that a state of motion is just as natural to a body as a state of rest, provided that no external cause acts upon it, or, in other words, that there is nothing in the inherent nature of a moving body to cause it to cease moving.

Here it is desirable that we should pause and reflect upon what is implied by the term Motion, and on the meaning of the phrases used to describe it. Motion means change of place, and it is obvious that in order to have a clear idea of the movement of any body, we must know the direction or line in which it is moving, and also the rate or velocity at which it moves. This velocity is often measured in ordinary life in miles per hour; thus an express train is said to go at the rate of forty miles an hour, meaning that if it continued moving for a whole hour at an unchanged speed it would rush over forty miles of railroad in that time; but that if it only travelled at that rate for a fraction of an hour it would move over a proportionate distance—two miles in three minutes, for example. It is more usual, in scientific language, to speak of velocity as measured in feet per second; for instance, sound travels through the air at the rate of about 1,100 feet per second.

The Laws of Motion were very carefully studied by that great philosopher, Sir Isaac Newton, who reduced them to three. The first one states that "a body at rest will continue at rest, and a body in motion will continue in motion, unless acted on by some external force;"

and it has already been discussed in connection
with Inertia. The second one states that "when
two or more forces act upon a body each produces its full effect, whether the body be at
rest or in motion." Simple illustrations of this
are seen in the fact that if a stone be dropped
from the window of a moving train it falls *on,
and not behind,* the carriage step, or, when let
fall from the top of the mast of a ship in motion,
it falls at the foot of the mast; in the feats
of circus performers, who, in jumping through
hoops, &c., simply jump *upwards* into the air,
and are carried *forwards* by the motion derived
from the horse; and in a boat which, propelled
straight forward by oars, while carried sideways
by the tide, moves in a line which is really the
result of the action of both these forces. The
third law states that "Action and Reaction are
equal and opposite;" but for present purposes
it need not be here discussed.

It may now be noted, however, with advantage, that although a body may be apparently
at rest, it may have a very decided *tendency to
move,* actual movement being prevented by some
counteracting tendency, a very slight diminution
in which may cause a vast movement to take
place. The case of heavy weights carefully
balanced, and then moved with a very slight
addition to them, illustrates this.

Before leaving the subject of Motion, it should

not be forgotten tl at a body may be affected by
motion in two ways; the first of these, viz., the
change of place of the whole mass of the body,
has been already considered; the other way is
by the movement among themselves of the small
particles of which the body is made up. This
kind of movement is known as Molecular Motion.
There are the strongest possible grounds for
believing that all matter, whether solid, liquid,
or gaseous, is made
up of exceedingly
small particles, invisi-
ble even with the aid
of the highest powers
of the microscope,
which are in a more
or less continual state

Fig. 4.

of movement among each other. When a blow,
or impulse of any kind, is sent through a solid or
liquid substance, it is passed on, as it were, from
particle to particle, somewhat in the manner represented on a comparatively gigantic scale in the
annexed figures (Fig. 4).

A good instance of this molecular motion,
or movement of the particles of an apparently

C

solid body amongst themselves, is afforded by
that curious alteration in the internal structure
of the iron in girders of bridges, &c., in consequence of which the iron becomes more brittle,
and the bridge therefore unsafe, after the lapse
of some years.   Illustrations of molecular motion
in liquids are afforded by the beautiful forms
assumed when they solidify—as for example,
the crystalline appearance of melted metals that

Fig. 5.

are allowed to cool slowly (as seen in ornamental
tin ware, &c.), or of different salts crystallising
from their solution in water; a very familiar
and a very beautiful instance is seen in the
exquisite crystals of snow, a few of the varieties
of which are seen in Fig. 5, and the same general
plan of structure exists also in ice, and is re-

vealed therein by passing a beam of light through
it, and throwing the image on a screen by a lens,
as shown in Fig. 6.

A very pretty ex-
periment illustrating the
molecular movement in
gases is as follows :—A
round porous pot (such
as is used in voltaic
batteries) is closed at
the open end with a
perforated cork, through
which a glass tube about
two feet long is inserted,
and the lower end of
this tube dips into co-
loured fluid (Fig. 7).
A bell-jar full of hydro-
gen gas (p. 70) is in-
verted over the porous
pot, when bubbles of air
will at once issue from
the mouth of the tube;
the reason of this is that
the molecules of hydro-
gen pass into the pot,
through its pores, faster
than the molecules of

Fig. 6.

air can pass out. If the bell-jar be now removed, the hydrogen will pass out faster than

c 2

the air can pass in again, a partial vacuum will
be formed, and the fluid will rise in the tube.

Although no one has ever yet seen a single
molecule, it is a remarkable fact that four distinct
lines of mathematical reasoning,
based upon as many different sets
of experiments, have led eminent men like Sir W. Thomson,
Clausius, and others, to the
conclusion that in any ordinary solid or liquid the mean
distance between the centres of
contiguous molecules is less than
the 1-5,000,000th and greater than
the 1-1,000,000,000th of a centimetre (2·5 centimetres = one inch).*
To form some idea of what this
implies, let the reader imagine a
solid sphere the size of a football
magnified to the size of the globe
of our earth, which is about 8,000 miles in diameter; if the molecules were magnified in the
same proportion, the structure would be more
coarse-grained than a globe of small shot, but less
coarse-grained than a globe of footballs.

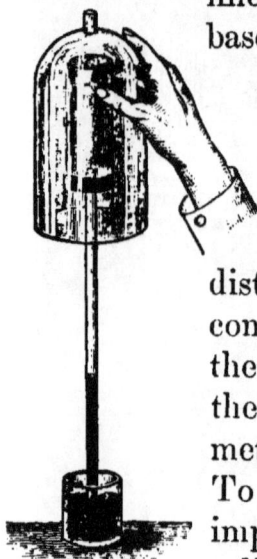

Since we have been considering the two kinds
of motion, molar and molecular, which affect

FIG. 7.

* For fuller information on this subject, the reader may consult
with advantage Sir W. Thomson's lecture, reported in three numbers of *Nature* for July, 1883.

matter, it may be convenient here to discuss
briefly the two forms of ordinary attraction
which affect matter, simply (so far as is at present known) as matter, *i.e.*, irrespective of such
attractions as those due to electricity, magnetism,
and chemical agency. The first, or the attraction between whole masses of bodies, is known
as Gravitation, while the second, which expresses
the attraction of molecules for each other, is
called Cohesion.

It is a universal physical law that every particle of the universe attracts every other particle
with a certain force, the amount of which depends on conditions to be presently explained.
If a body (such as a stone or a rifle bullet) be
projected into the air, experience tells us that
it will almost immediately return to the surface
of the earth, in consequence of the attractive
force exerted by the earth upon it. It is equally
true, however, that, in an infinitesimal degree,
the earth falls towards the stone. Very delicate
experiments have been made in a laboratory with
two large masses of lead, accurately balanced
and free to move, which proved the fact of the
attractive force that they exerted on each other.
A good illustration of the attraction of a large
for a small mass of matter is seen in the fact
that a plumb-line in the neighbourhood of large
mountain masses does not hang perfectly vertical,
the " bob " being attracted towards the moun-

tain, as was clearly shown by experiments made
near Schehallion, an isolated mountain in Scot-
land. Astronomers tell us how gravitation ex-
tends its power to the most remote parts of the
heavens; how the moon and the earth gravitate
towards each other, and both together towards
the sun; how the whole solar system is gravi-
tating in a given direction among the stars;
how double stars are systems of suns, revolving
round each other under the action of the same
force, each of which has probably its own system
of planets, all moving in accordance with the
same law. If we confine our attention to this
earth and its satellite, we shall see that to gravitation in great part are due, in one sense the
rise and fall of the tides, all the phenomena of
wind and rain, all the mechanical energy derivable
from falling water (in the form of torrents, so
apparently irresistible), and, in fact, the regu-
lation of the whole of the existing fabric of the
earth's surface. Were it not for the attraction
of gravitation, the speed at which the earth is
rotating on its axis would send every loose thing
upon its surface flying into space; if we jumped
into the air, or went up in a balloon, we should
have no means of returning to the earth's surface. Thoughts like these illustrate the truth of
the remark that "gravitation is the force which
keeps the universe going." It is by virtue of
gravitation that matter possesses weight; for

the weight of any thing is the expression of the force with which it tends towards the earth, and, as we have seen, depends upon the mass, or quantity of matter, in the body. Hence it is evident, first, that weight is a particular case of the universal law of gravitation, and, secondly, that the force of gravitation between two bodies depends upon their mass. It remains to enquire what effect distance has upon the force also. Experiment shows that if the distance between two bodies be doubled, the force of attrac tion between them will only be one-fourth as great; if the distance be trebled, one-ninth as great; if quadrupled, one-sixteenth as great, and so on; this is expressed in general terms by saying that the force of gravitation varies inversely as the square of the distance. Other physical phenomena follow the same rule, *e.g.*, the distribution of light from any light-source. Another illustration of the law of squares is seen in the phenomena of falling bodies; a stone falls towards the earth 16 feet in the first second, 64 feet (or 16 multiplied by the square of 2) in 2 seconds, 144 feet (or 16×3^2) in 3 seconds, and so on. If the resistance of the air were removed all bodies would fall towards the earth at the same rate, irrespective of their comparative weight. This may be roughly shown by placing upon a penny a flat disc of paper the same size as the coin, and letting both fall together, when they will be found to reach the ground at the same instant.

In all bodies there is some point at which its whole weight may be considered as concentrated, and about which it will balance in all positions; it is called the centre of gravity. When a body is free to move it takes up such a position that its centre of gravity occupies the lowest possible point, hence by hanging up an irregularly shaped body in two positions (Fig. 8), and noticing where vertical lines from the points of support intersect each other, its centre of gravity may be found. Many remarkable feats of balancing, of recovery of equilibrium, &c., depend upon the position of the centre of gravity of the performer or of the thing balanced. The general rule governing these cases is that the equilibrium will be maintained as long as a vertical line drawn downwards from the centre of gravity of the body, falls within the base-line of the support (Figs. 9 and 10). The celebrated leaning tower of Pisa is a standing instance of this.

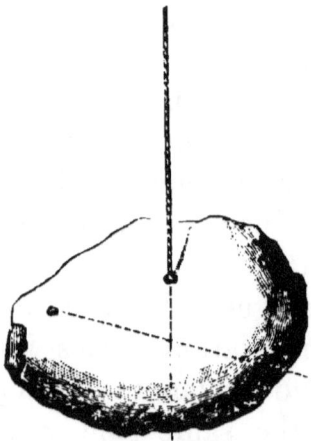

FIG. 8.

The attraction of cohesion differs from that of gravitation, inasmuch as it is only exerted between very minute particles of matter, and at exceedingly small distances; it is, however, that which binds together the various particles of a body, and resists our attempts to break it; on the other hand, when the body is once bro-

Fig. 9.

ken, and the particles separated by a measurable distance, it is very difficult to get them to adhere to-gether again.* It is obvious, therefore, that this molecu-lar attraction acts very powerfully

Fig. 10.

through a certain small distance, but disappears

---

* A notable instance to the contrary is afforded by lead; a bullet may be cut in half, and the two clean surfaces will adhere strongly if pressed together. Adhesion is the name sometimes given to this form of cohesion.

entirely when the distance becomes perceptible. Cohesion is strongest in solids, in liquids
it is much diminished, and in gases it may
be said to vanish altogether. Reference was
made to the conditions of material life without gravitation; a similar line of thought with
regard to cohesion shows that if that attraction
suddenly ceased to be exerted, our houses, our
own bodies, and, in fact, the whole surface of the
earth, would be instantly resolved into a heap
of dust! This is well illustrated by the so-
called Prince Rupert's drops, formed by allowing
drops of melted glass to fall into water, which
assume, on cooling, the form of a pear-shaped
crescent. If the conditions of equilibrium (*i.e.*,
the balancing of the forces) of cohesion are disturbed by breaking off the minutest fragment
from the small end, the whole mass (the thick
part of which will resist severe blows from a
hammer) will suddenly fall into fine powder.

Having thus disposed of some of the more
important properties of Matter, simply as matter,
let us proceed to inquire a little more in detail
into what we have ventured to term its moods, *i.e.*,
to discuss the states and forms of Energy, which
forms the special subject of this book.

Energy may exist in two states, that of
motion, and that of repose. It is obvious that
any moving body possesses energy, or the power
of doing work, but it is not quite so clear at first

sight how there can be energy in repose. In order to understand it, let us recur again to the analogy of the energetic man; he may be very quiet, and yet be able to do a great deal of work when he chooses to set about it. Energy in repose is in many cases due to its position, as may be seen from this example:—Imagine a mill or factory driven by a water-wheel, and near this mill two ponds of water of equal capacity, one below the level of the wheel, the other at a considerable elevation above it. It is obvious that, as far as the mill is concerned, there is no work at all to be got out of the lower pond, while the fall of water from the upper pond will drive the wheel. It should be carefully noted that the work is done by the passage of the water from a higher to a lower level, and this will help us to understand later on, how, in doing work, energy passes from a higher to a lower grade. Examples, then, of energy due to position, or, as it is called, *potential energy*, are to be found (mechanically) in a raised weight, a stretched spring, and a "head" of water; chemically, in coal, gunpowder, corn, and other foods, &c. (Chaps. III. and VI.); electrically, in voltaic batteries, and in accumulators or storage batteries (Chap. IV.).

A little consideration will show that all potential energy must ultimately be converted into energy of motion (called scientifically *kinetic energy*), whether that motion be molar (*i.e.*, of the mass), as in the case of mechanical motion, or one of the various forms of molecular motion.

The following table gives the various supplies of natural energy and their sources, it being borne in mind that the words are used in a wide sense—fuel and chemical affinity, for example, being intended to include the burning of zinc in a voltaic battery (pp. 102, 119) as well as of fuel in the air.

Supplies of Natural Energy.

Potential.	*Kinetic.*
(1) Fuel.	(1) Winds.
(2) Food of animals.	(2) Water and ocean currents.
(3) Ordinary water-power.	(3) Volcanoes and hot-springs.
(4) Tidal water-power.	

The sources of these supplies are :—

   (1) Chemical affinity.   (2) Solar Energy.   (3) Energy of the earth's rotation on its axis.   (4) Internal heat of the earth.

These energies may be conveniently distributed into three groups, as follows :—(1) Visible energies, including all cases of visible motion, such as those produced by gravitation, &c; (2) heat energies, in which the motion is molecular (Chap. II.); and (3) chemical and electrical energies, in which the motion is rather that of atoms than of molecules (*see* Chaps. III. and IV.) It must not be imagined, however, that each of these groups or systems is complete in itself and has no sort of connection with its neighbours, for the various forms of energy are inseparably intermingled with each other.

Energy, then, being the power of doing work, let us next enquire how that work is measured, a point of the utmost importance, as to which clear ideas are essential. When a man raises a weight from the ground, against the action of gravity, he does a certain amount of work; if the number of pounds in the weight raised be multiplied by the number of feet through which it is raised, a number will be obtained which expresses the measure of that work; these numbers are called foot-pounds, *and foot-pounds are the standard in which all work is measured scientifically.* One foot-pound, then, is 1 lb. raised 1 foot; 100 foot-pounds is either 1 lb. raised 100 feet, or 100 lbs. raised 1 foot, or 10 lbs. raised 10 feet, and so on.

The "horse-power" is the measure of the rate of work most commonly used, but it is obviously an unscientific one, because the strength of horses varies so much. What is now recognised as one horse-power is 550 foot-pounds of work per second, or 33,000 foot-pounds per minute.

In the metrical system, to which reference has been made, the standards taken are the kilogramme (or 1,000 grammes, about 1·1 lb.) and the metre (39·37 inches), and hence the unit-measure of work on the continent, and among many English scientists, is the kilogram-metre, which is equal to about $7\frac{1}{4}$ foot-pounds.

Since the various forms of energy are inter-changeable, this work-measure can be applied to any one of them, whether it be heat, or electricity, or any other form, and so it comes to pass, for example, that when electrical energy is applied to produce mechanical motion (Chap. V.), and is distributed over our towns for this purpose, the amount supplied will be measured, and can be expressed in foot-pounds, by meters constructed for the purpose.

It is a matter of common experience that the energy of a moving body depends upon, or varies with, its mass. The relation that the velocity of a body bears to its energy is expressed by the law of squares (p. 23); i.e., if the velocity be doubled, the energy is quadrupled, three times the velocity gives nine times the energy, and so on.

Having now cleared the ground somewhat, and gained some exact ideas on the subject of matter and its moods, let us enquire into the truth of the statement so often insisted on, that energy is never destroyed, but only transmuted, or changed in form; and, to begin with, let us try to find an answer to the enquiry, What becomes of the energy of mechanical motion when it is retarded, or altogether arrested ? It is a matter within the common experience of

almost every one, that any attempt to stop such
motion, to "put the brake on," results in the
production of heat, and it will be the object of
the next chapter to point out the relations between
mechanical energy and heat energy.

CHAPTER II.

LET us now consider some instances of the statement that whenever mechanical energy is retarded or destroyed, heat makes its appearance. All who have anything to do with machinery are familiar with the fact in the form of hot bearings, which need to be lubricated with oil (or water) in order to diminish the friction; in the absence of that appliance of civilised life, the lucifer match (which is itself ignited by the heat produced when it is forcibly drawn over a rough surface); the untutored savage spends a great deal of energy in producing fire by rubbing together two pieces of wood; a clever blacksmith can heat a large nail red-hot by simply hammering it upon his anvil; when a cannon-shot strikes a target, its energy of motion is at once destroyed, but both shot and target become very hot, and fragments of the metal are often heated sufficiently to glow perceptibly in diffused daylight. The annexed engraving (Fig. 11) represents a Shoeburyness target struck with an amount of energy expressed by about 2,200 *foot-tons*.

The first person to whom science was in-

debted for proving clearly the relationship between mechanical and heat energy, was Count
Rumford, who, on January 25th, 1798, read an
elaborate paper to the Royal Society, entitled
" An Enquiry concerning the Source of Heat
which is excited by Friction." In it he detailed

Fig. 11.

experiments showing the large amount of heat
produced in boring brass guns by horse-power
in the military arsenal at Munich, during which
the gun itself became hot enough to boil large
quantities of water, and the metallic chips were
intensely heated. These experiments, taken in
connection with the very obvious fact that a
hot body does not weigh more (or less) than the
same body when cold, effectually disposed of

D

the notion that heat was a material substance. " What is heat ? " he asked. " Is there any such thing as an *igneous fluid?* Is there anything that, with propriety, can be called caloric?" and, after reviewing the whole series of experiments, he concluded, "It appears to me extremely difficult, if not impossible, to form any distinct idea of anything capable of being excited and communicated in these experiments, except it be MOTION." There is, as will be seen in the sequel, good reason to

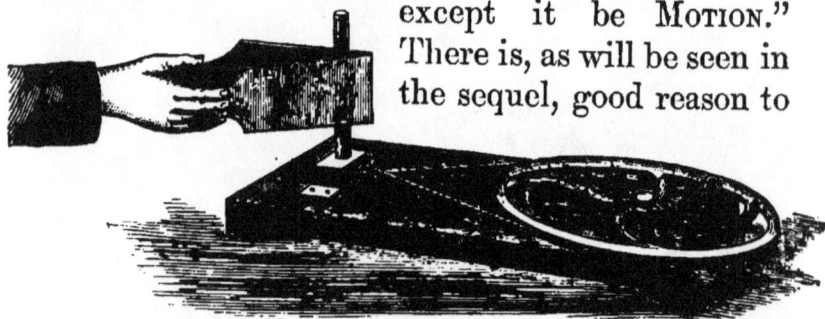

FIG. 12.

believe that when motion disappears in this way, it is changed from the motion of the whole mass of the body into the motion of its particles, or, in other words, that the molar motion becomes molecular. Count Rumford's experiment may be illustrated on a small scale by the apparatus represented in Fig. 12, where a tube four inches long, and three-quarters of an inch in diameter, containing water, and corked, is rapidly rotated, while it is gently squeezed with a broad pair of wooden tongs. The heat thus developed speedily boils the water and blows the cork out.

To Dr. Joule, of Manchester, belongs the credit of first expressing in numbers this relationship between heat and work, by a long series of laborious researches, extending from 1843 to 1849, and his experiments were of this kind. He took a closed vessel B, Fig. 13, containing water, in which a paddle fixed on an axis was caused to rotate by gearing connected

F<small>IG</small>. 13.

with falling weights E, F; and as the amount of this weight was known, and also the distance through which it fell, the *work done* could be calculated in foot-pounds; the friction of the paddles against the water heated the whole contents of the box, and thus Dr. Joule established the fact that the mechanical work represented by 772 foot-pounds would, when converted into heat, raise the temperature of one pound of water by one degree of Fahren-
D 2

heit's thermometer (the ordinary scale in use in
England).   This number, then, is known as the
mechanical equivalent of heat.   Conversely, the
amount of heat necessary to raise 1 lb. of water
1° Fahr. would, if all applied mechanically, raise
772 lbs. 1 foot high, nearly equal to 1½ horse-
power for 1 second.

The heat generated by the collision of a falling body with the earth depends on the height
from which it falls, and as that height is proportional to the square of the velocity (p. 23),
it follows that the heat generated increases as
the square of the velocity; hence, if we double
the velocity of a projectile, we increase four-
fold the heat generated when that motion is
destroyed, and so on.   A velocity of about 1,400
feet per second in a rifle bullet would, when
it struck the target, raise its own temperature
nearly 1,100° Fahr. (if no heat were absorbed
by the target), which would melt a portion
of the lead.   Calculating on this principle
it has been shown that if the earth were
suddenly stopped in her motion through space,
as much heat would be generated as would
be developed by the combustion of fourteen
globes of solid carbon, each as large as the
earth; and that if the earth fell into the sun
the heat thus produced would be equal to that
of the combustion of 5,600 such worlds.   Specula-
tions upon how the vast amount of the sun's

energy * is maintained have led to the suggestion that it may be kept up, in part at any rate, by the constant showering down of solid meteoric matter upon its surface!

It is a very curious fact, and one that has an important bearing upon the economy of nature, that equal weights of various bodies require different quantities of heat to bring them to the same temperature, or, in other words, that bodies vary very much in their capacity for heat. If 1 lb. of water at 60° Fahr. be mixed with 1 lb. of water at 212° Fahr. (its boiling-point) the temperature of the mixture will be found to be half-way between the two, or 136°; but if 1 lb. of mercury at 212° be used instead, the temperature of the mixture will only be about 65°. Water has a very great capacity for heat; the same amount of heat that will raise the temperature of 1 lb. of water 1° Fahr. will raise 9 lbs. of iron, 11 lbs. of zinc, and 30 lbs. of mercury by the same amount. This difference in capacity for heat may be also thus shown experimentally; balls of different metals, iron, lead, bismuth, tin, and copper, are heated in oil

---

* It is difficult to convey any idea of this. The amount of heat received by the earth from the sun in one year would liquefy a layer of ice 100 feet thick all over it, and yet this is only the 1-2,138,000,000th part of the total heat emitted. Sir W. Thomson has lately put it in another form (*Nature* for January 18, 1883), and expresses the radiant energy of the sun as equal to 7,000 horse-power per square foot, or 50 horse-power per square inch, every second of time, from the whole of his vast surface.

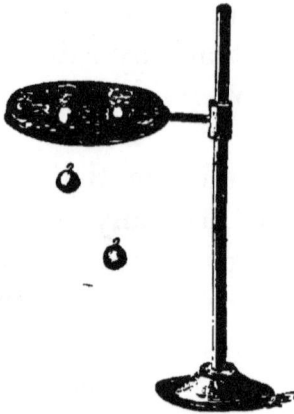

FIG, 14.

to about 350° Fahr. and laid upon a cake of wax (Fig. 14). The iron and copper balls will work themselves through first, the tin will follow, while the lead and bismuth scarcely sink more than half the depth of the wax.

The following table indicates in figures the relative capacity for heat, or *specific heat*, of various substances :—

Substances.						Specific Heat.
Water	.	.	.	.	.	1·000
Turpentine	.	.	.	.	.	0·426
Air	.	.	.	.	.	0·237
Glass	.	.	.	.	.	0·198
Iron	.	.	.	.	.	0·114
Copper	.	.	.	.	.	0·095
Tin	.	.	.	.	.	0·056
Mercury	.	.	.	.	.	0·033
Lead	.	.	.	.	.	0·031

The specific heat of water is nearly the greatest of all known substances, hydrogen, and a certain mixture of alcohol and water, being the only exceptions. Comparing equal *weights* the relation of the specific heat of water to that of air is as 1 to 0·237, or 4·2 to 1; but as water is 770 times as heavy as air, if equal measures are compared it will be seen that the proportion is as 770 × 4·2, or 3,234 to 1. Hence a cubic foot of water in cooling 1° will warm by the same amount 3,234 cubic feet of air. A little reflection will show the . bearing of this fact upon the influences of the sea on climate, and the reason of the comparative absence of extremes of heat and cold in an island climate ; it will presently be shown also how the system of circulation of the waters of the great oceans, from the equator to the poles, and from the lowest depths to the surface, demonstrated by such researches as those of H.M.S. *Challenger*, assists in moderating the severity of the land climate in various parts of the earth, and notably in the western shores of Europe.

We have now to inquire into the mode in which heat passes through bodies. Experiment shows us that substances vary very much in their power of allowing this kind of molecular motion to pass through them, or, in other words, of conducting heat. A simple illustration will show this : place two spoons, one of pure silver, the other of German silver, in the same vessel of hot water ; in a few moments the free end of the silver one will be much hotter than its neighbour, and if the experiment be repeated, and fragments of phosphorus placed on the ends, that on the end of the silver spoon will fuse and catch fire, while the other piece will be unaffected.

Of all solid bodies, metals (except bismuth) conduct heat best; next in order come stone, glass, and marble, wood-charcoal, and animal and vegetable tissues. The following table shows the relative conducting powers of metals for heat and for electricity (pp. 106–7), both of which we have good reason to believe are different kinds of molecular motion. It will be noticed how alike they are :—

Substance.	Conductivity for Heat.	Conductivity for Electricity.
Silver	100	100
Copper	74	73
Gold	53	59
Brass	24	22
Tin	15	23
Iron	12	13
Lead	9	11
Platinum	8	10
German Silver	6	6
Bismuth	2	2

Advantage is taken of the unequal conductivity of solids in many ways. Tools and metal utensils are furnished with non-conducting handles of wood or ivory: a wooden house is cooler in summer and warmer in winter than a stone one; ice is packed in sawdust or flannel to prevent its melting; steam boilers and pipes are (or should be) covered with felt. The natural clothing of the animal creation, and its application to the needs of human life, depend upon the same circumstance; and the effect is heightened in this case by the fact that the air entangled among the fibres, feathers, &c., is an almost perfect non-conductor, so long as it is at rest.

It should be noticed in this connection that the sensation of cold is really due to an abstraction of heat from our own bodies. The temperature of the blood is about 98° Fahr., and when the hand is laid upon any substance at the air-temperature, it feels more or less cold, according to the rate at which heat passes from the hand to it, *i.e.*, according to the conductivity of the substance. In the severe winter of the Northern United States it is absolutely necessary for all who have to handle metals in the open air to wear gloves; the metal is not colder than the surrounding air, but, being a good conductor, it robs the human body of heat very rapidly, and will produce blisters upon the naked skin.

The experiments which have been made to measure the conductivity of liquids and gases, prove that it is very slight indeed. Heat is, however, rapidly transmitted through these media by actual transport of the heated parts, a process to which the term *convection* (*i.e.*, conveyance, carrying,) is given; it depends on the fact that (as will be immediately shown) the heated portions of the liquid expand, and thus become specifically lighter, when the balance is therefore disturbed, and motion throughout the mass ensues. The annexed cut (Fig. 15) shows the convection currents set up in water when

heated from below. Numerous examples of convection on a grand scale in nature will readily
occur to the thoughtful mind: the cooling of
large masses of fresh water in winter, the movements of air in all systems of ventilation, the
trade-winds, produced by the sun heating the air
at the equator, and, indeed, all
winds. The investigations into
the temperature of the sea at
various depths, made on board
H.M.S. *Porcupine, Challenger,* &c.,
show that in the great oceanic
basins there is a constant circulation of their waters, both
vertical and horizontal; that the
upper layers are moving from the
equator towards the poles, while
the lower layers are gradually
creeping along the floor of the
oceans from the poles towards the equator; both
upper and lower currents, however, being in-
fluenced in their direction, just as the trade-
winds are, by the rotation of the earth upon its
axis, the northerly and upper current in the
northern hemisphere being turned in an easterly
direction, and the southerly and lower current
in the same hemisphere taking a westerly set.
There is good reason to believe that the maintaining cause of this grand circulatory system is to be
sought rather in polar cold than in equatorial heat.

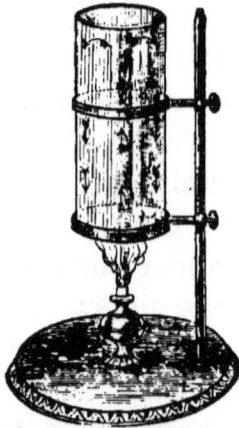

FIG. 15.

The interchange of heat between bodies not in contact, takes place by the process known as radiation. It is in this way that any heated body cools in the air—that we are sensible of heat when we approach a fire—that the heat of the sun reaches the earth and the various planets. Radiant heat passes through space in straight lines at the same speed as light, and can be reflected by mirrors and refracted by lenses in precisely the same manner, a familiar illustration of which is seen in the so-called burning-glass.

The nature of the surface of a body has a very great influence upon the rate at which it loses heat by radiation, and since those surfaces which radiate heat best, also absorb it most readily, these differences have very important bearings upon natural phenomena, and, therefore, indirectly upon human life. Moreover, the power of a surface to reflect heat is the complement of its power to radiate or absorb it, as is seen by the following table, the totals of the two numbers opposite any substance being 100 in all cases :—

	Radiating powers.	Reflecting powers.
Lamp black . . .	100	0
Glass	90	10
Steel	17	83
Platinum . . .	17	·83
Polished brass . . .	7	93
Red copper . . .	7	93
Gold	5	95
Polished silver . . .	3	97

The properties of radiant heat have been made a special subject of experimental study by Prof. Tyndall, who has shown how closely allied they are to those of the other form of energy which we recognise as light. The term radiant energy may, perhaps, be used to include them both, and a very pretty illustration of its effects is seen in the Radiometer of Mr. Crookes, in which four small mica vanes, blackened on one side, are mounted upon a pivot, and the whole arrangement is placed in a globe from which air is then almost completely exhausted (Fig. 16). When radiant energy, as the light of a candle, is allowed to fall on the globe, it is absorbed by the black sides of the vanes, and the molecular motion thus set up among the particles of gas still left inside the globe is at once transformed into visible motion, and the vanes rotate rapidly, with a speed varying partly with the intensity of the light.

Fig. 16.

It must not be hastily concluded, however, that radiant heat and light are identical, although they are propagated in the same way, viz., by wave-motion in that ether, which, according to the "undulatory theory" now generally accepted, pervades all space. The length of these waves can be measured, and those which affect the sense of touch as heat are much longer than those which affect the eye as light. Further, many substances, glass for example, are quite transparent to light, but opaque to radiant heat; while others, such as iodine in solution, are absolutely opaque to light, but permit radiant heat to pass with the greatest ease. The presence of more or less moisture in the air has a most important influence on the passage through it of radiant heat. Moreover, it must be borne in mind that the passage of radiant heat, as such, through any medium *does not heat it at all;* radiant heat only becomes sensible heat when the waves by which it is propagated are absorbed, instead of being either reflected or transmitted.

We now pass to the consideration of the effects of heat upon Matter, one of the first and most obvious of which is that, with one or two exceptions, all bodies expand when heated, and contract when cooled. In the case of solids this expansion takes place in the three directions of length, breadth and thickness. The linear expansion of a rod by heat has been used in an instrument called a pyrometer to measure very high temperatures, in which an expanding rod presses against the short end of a bent lever, the long end of which moves a pointer upon a scale. Solids vary much in the amount of

their expansion under the same rise of temperature, and the mechanical energy of this molecular motion under changes of temperature, is
very great; an iron bar, for example, one inch
square, cooled through 80° Fahr., contracts with
a pull of fifty tons. Advantage has been sometimes taken of this fact to restore to the per-
pendicular the bulging walls of a building, iron
tie-rods being placed across, and while they
were red-hot the nuts on the ends of the rods
outside the walls were screwed up tight; on
cooling, the contraction of the rods drew the
walls together. The practice of "shrinking"
tyres on wheels is an example of the force of
contraction.

As a general rule liquids expand more than
solids, but also vary much among themselves.
The indications of the thermometer depend
upon the expansion of liquids by heat. A glass
bulb, from which projects a long fine tube, is
filled with either mercury or alcohol; in order
to graduate it, it is plunged first into melting ice,
and a mark is made on the stem where the column
of fluid becomes stationary; the same process is
then repeated with boiling water. The interval
between these two points is variously divided;
the scientific division, also very largely used on
the continent, is into 100 parts, the lowest mark
(melting ice) being called 0°, and the highest
(boiling water) 100°. This is called the centi-

grade, or "hundred steps" scale, and ought to be universally adopted. In Northern Europe the interval is divided into 80°, melting ice being 0°, a scale known as Reaumur's. In English-speaking countries the scale is that of Fahrenheit, in which the interval is divided into 180 parts; in this case, however, the 0°, or zero-point, is not the tempera-ture of melt-ing ice, but a point as much below that as cor-responds in distance on the scale to 32 of these 180 parts. Hence, on

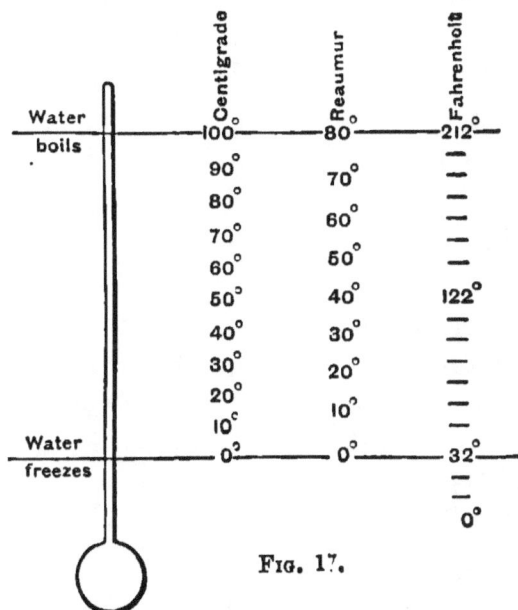

Fig. 17.

Fahrenheit's scale, the "freezing-point," which is the zero of the two other scales, is marked 32°, and the boiling-point of water is 32° + 180°, or 212°. Since, therefore, the same in-terval, *i.e.*, that between the temperatures of melting ice and of boiling water, is variously divided (Fig. 17) into 100°, 80°, and 180°, the proportions of degrees on the Centigrade, Reau-

mur, and Fahrenheit scales are as 100 to 80 to 180,
or as 5 to 4 to 9, or in other words, 5° C. =
4° R. = 9° Fahr. The temperatures of melting
ice and of boiling water being the two fixed
points, all degrees above and below those are
obtained by simply prolonging the scale in either
direction.

The most notable exception to the law of
expansion of liquids by heat is the case of fresh
water, which is at its greatest density (*i.e.*, is
heaviest in proportion to its bulk) at 4° C.
(39° Fahr.). Heated above, or cooled below,
this point, it expands, and when it is converted
into ice the expansion is sudden and considerable, the ice, as is well known, floating on the
surface. Were it not for this remarkable exception to the general law, fresh-water lakes would
become solid masses of ice in severe winters, all
animal life therein would probably perish, and
the climate in their neighbourhood would become quite rigorous.

It is a somewhat remarkable fact that, in
the absence of experiments, physicists somewhat
hastily assumed that sea-water followed the same
rule, and hence predicted that "in all deep seas
a temperature of 4° C. (39° Fahr.) would be
found to prevail."* Experiment shows, however,
that sea-water continues to contract down to its
freezing-point, and the deep-sea temperature

* Sir J. Herschel's "Physical Geography," 1861.

observations before referred to confirm the fact, and demonstrate the important consequences that follow from it.

The linear expansion of metals heated between the freezing and boiling points of water, varies from about one to three parts in 1,000. Water similarly treated undergoes a total increase in volume of 43·15, *i.e.*, 1,000 gallons would become 1,043½. Three cubic feet of air, or of almost any gas, heated under the same circumstances, would become four, or, more exactly, 1,000 cubic feet would become 1,367, provided that the pressure were unaltered.

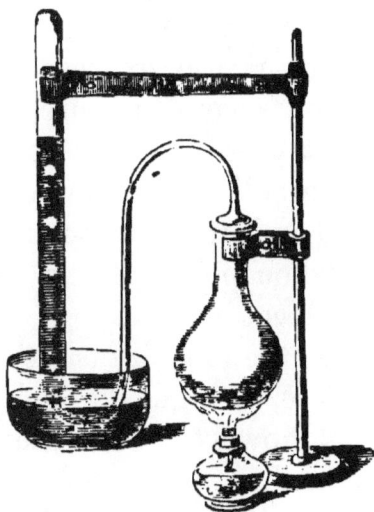

FIG. 18.

The expansion of gases by heat may be readily shown by heating some in a vessel (Fig. 18) provided with a tube, the open end of which dips under a vessel previously closed at one end, filled with water, and inverted. The ascent of a fire-balloon, the ventilation of mines, the ascending currents of air which produce winds (and which under various conditions of moisture cause clouds also), are all illustrations of the effects of alterations in the density or specific

E

gravity (p. 10) of air, produced by the expansion caused by heat.

Having now considered the alterations in the mass of a body produced by heat, let us consider more closely its effect upon the molecules of which that body is composed. It was pointed out in the last chapter, water being taken as an illustration, that whether any substance was in the solid, liquid, or gaseous form depended in great measure upon the temperature and pressure to which it was exposed. By the use of intense cold and severe pressure, even the so-called permanent gases—such as oxygen and hydrogen— have recently been condensed to liquids. On the other hand, the intense heat of the arc between the carbon points of the electric light (Chap. V.) is sufficient not only to melt, but to turn into vapour, the most infusible metals. Heat-energy, therefore, changes the molecular state of matter.

If, however, the change be more closely examined, it will be found that in the passage of any substance from the solid to the liquid, or from the liquid to the gaseous state, an enormous quantity of heat disappears; and that any change in the reverse direction is always accompanied by the apparent production of heat. For example, if 1 lb. of water at 0° C. (32° Fahr.), and 1 lb. of water at 77·8° C. (172° Fahr.) be mixed together, the result will be 2 lbs. of water at 39° C. (102° Fahr.). On the other hand, the temperature of a mixture of 1 lb. of *ice* at 0° C. and 1 lb. of *water* at 77·8º C., will be found to be only 0° C. What has become of this heat? The only difference between the two experiments is, that in the first case *liquid* water was used, and in the second ice, or *solid* water. In the older books the phenomena was said to be explained by saying that the heat which thus disappeared "became latent," and the phrase is still in use to *express* the fact, but it does not *explain* it. Why does the heat become latent? The true explanation, upon the principle of the conservation of energy, is that *it is used up in overcoming* the cohesive force of the molecules of water, and is thus transformed into a kind of energy of position.

Again, it can be shown experimentally that 1 lb. of water at 100° C. in being turned into steam, absorbs enough heat to raise 537 lbs. one degree in temperature, and *yet the steam is no hotter*. In this case also the heat has "conferred potential energy upon the atoms," as any attempt to make the experiment in a confined space will immediately render evident! A cubic inch of water produces nearly a cubic foot of steam.

It may be stated generally, then, that change of state in the direction of solid to gas is accompanied by the absorption of heat, or, in other words, the production of cold. Freezing mixtures, in which certain substances—snow and salt, for example—rapidly liquefy when brought

E 2

into contact, depend upon this principle, as well as
all those freezing machines which owe their action
to the rapid vaporisation of some volatile liquid,
as ether, liquid sulphurous acid, or solution of
ammonia. In all these cases the heat which is
thus abstracted, reappears in that part of the
machine devoted to the condensation of the
vapour.

In a similar way, whenever work is spent
upon a gas, as it is when air (or gas) is compressed by mechanical means, heat is evolved,
and when that gas is allowed to expand again by
the removal of the pressure, heat is absorbed, or,
in other words, cold is produced. The heat produced by the compression of air may be shown
experimentally by placing a piece of dry tinder
under the piston of an air-syringe, closing the
mouth of the cylinder, and smartly driving the
piston to the bottom of it; the heat thus evolved
will ignite the tinder, as may be seen when the
piston is withdrawn.

During the last few years several very success-
ful attempts have been made by various practical
inventors, such as Bell-Coleman, Hargreaves, and
others, to take advantage of the production of
cold when compressed air is allowed to expand,
and to construct cooling machines upon this principle. Such machines (Fig. 19) are now extensively used for freezing meat, and for maintaining so low a temperature in the chambers in

which it is stored, that several cargoes of fresh
meat have lately been brought to England from
Australia and New Zealand in such good condition as to be indistinguishable from home-
grown meat. The air is compressed by pumps,
operated by a small steam-engine, and confined
in strong reservoirs, round which sea-water is

Fig. .

allowed to flow, in order to cool the compressed
air to the surface-temperature of the ocean, after
which it is allowed to expand into the frozen
meat-chamber, which is of course kept closed
(except as to the air entrances and exits) and
surrounded with non-conducting material. The
temperature of this chamber is controlled by
regulating the volume of air passing into it, the
quantity required being naturally dependent

upon its temperature, and this again upon the
pressure from which it is allowed to expand. A
very obvious advantage of this process is that no
" chemicals " are employed in it by which the
flavour of the meat can possibly be affected.

Let us now consider that machine which
effects the transformation of heat into work for
practical purposes—the Steam-engine. It is well
known that the work, or motion, is produced
by the expansive force of steam, which is admitted alternately on either side of a piston
fitted tightly in a closed cylinder in which it
moves to and fro, and that when it has done
its work the steam passes either into the air (as
in locomotives and other "high-pressure" engines), or into a cold chamber or condenser (as
in the ordinary marine and stationary engine).
It will probably surprise many, however, to be
told how very imperfect a machine even the
most modern type of engine is, not more than
eighteen per cent., or less than one-fifth, of the
energy generated by the combustion of the fuel
being given back as mechanical work. In fact,
a no less eminent authority than Sir W. Armstrong has stated that, for practical purposes,
if the whole potential energy of the coal be
divided into ten parts we should find that two
went up the chimney, one was lost by radiation and friction, *only one was turned into work*,
and the remaining six were wasted! If it were

not that coal is so cheap, and every other form of potential energy that we can buy is so dear, we should find the steam-engine very expensive to use. An attempt will now be made to show how this comes about, and why no very great improvement in the steam-engine is to be expected.

There is one condition which must be rigidly fulfilled in order to get mechanical work out of heat—there must be a difference of temperature; and the heat must pass from a body of high temperature to one of low. An analogy may here be drawn with the case of water, out of which no work can be got unless it flows from a higher to a lower level (p. 27). What difference in temperature, then, is it possible to maintain in practice between the boiler and the condenser of the steam-engine? In giving an answer to this question we must take into account not merely the thermometric differences, but the absolute quantity of heat, or of heat-units, in each, and to do this we must enquire what is the *absolute zero of temperature.* The law of expansion of gases by heat tells us that gases expand $\frac{1}{273}$ of their volume for every increment of 1° C. in temperature between 0° and 100° C., so that at + 273° C., the elastic force of a gas is double what it is at 0° C. Supposing the same law to hold good in the other direction, at –273°, *i.e.,* 273° below zero, the gas would have no

elastic force at all. This point, then, –273° C.,
or –461° Fahr., may be considered as probably
the *absolute zero* of temperature, although it has
never been actually reached. This number of
degrees, therefore, must be added to the thermometric degrees in any such calculation. Let us
assume the case of an ordinary engine with
steam at three and a half atmospheres pressure,
or 53 lbs. per square inch. The temperature of
this is 300° Fahr. The condenser cannot practically be kept below 110° Fahr. Hence we
have :—

Heat in boiler .   . 300° + 461° = 761 units of heat.
Heat in condenser . 110° + 461° = 571   „   „   „

Difference, available for work   . 190   „   „   „

Or only one-fourth of the total energy (190
parts out of 761) is available for the production
of motion, even supposing that there were no
other sources of loss, such as friction, radiation,
&c. How much more perfect a machine in this
respect is nature's engine, *i.e.*, the human body,
in which energy is derived from the combustion
of food, will be seen in Chap. VI. (p. 189).

It has been pointed out by Thomson, that
although work can be transformed into heat with
the greatest ease, there is no process known by
which all the heat can be changed back again into
work ; that, in fact, the process is not a reversible
one. The consequence is that the mechanical

energy of the universe is daily becoming more and more changed into heat, that heat being of a low grade, and (as we have seen in the case of the steam-engine) inconvertible. Hence it is conceivable that a time may ultimately arrive when "the universe will become an equally heated mass, utterly worthless as far as the production of work is concerned, since such production depends upon difference of temperature. Although therefore, in a strictly mechanical sense, there is a conservation of energy, yet, as regards usefulness or fitness for living beings, the energy of the universe is in process of deterioration. Universally diffused heat forms what we may call the great waste-heap of the universe, and this is growing larger year by year. . . . We are led to look to a beginning in which the particles of matter were in a diffuse chaotic state, but endowed with the power of gravitation, and we are led to look to an end in which the whole universe will be one equally-heated inert mass, and from which everything like life, or motion, or beauty, will have utterly gone away."* This is the doctrine known under the name of the " Dissipation of Energy," and, although very suggestive, its consideration should be entered upon with the recollection that it applies solely to the physical universe, or rather to such portions of it as our senses can appreciate.

* " Conservation of Energy," by Balfour Stewart, p. 153.

Reference has already been made to the intimate association of heat and light, and it may be well here to point out that all bodies when heated sufficiently give out light, and that the colour of the light in the case of solids and liquids (melted metals, for example) depends upon the temperature to which the body is heated. The phrases in common use, such as red-

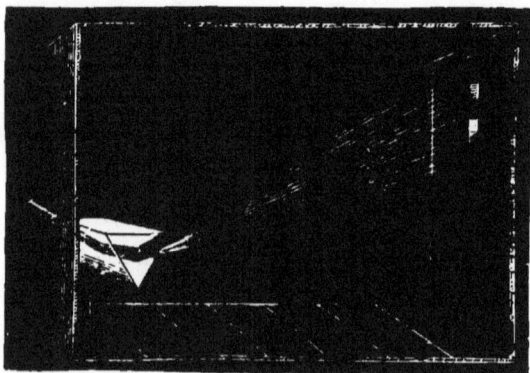

Fig. 20.

hot, white-hot, &c., applied to metals, recognise this fact. A very refined method of examining or analy-

sing the colour of light, is to pass a beam of it through a triangular piece of glass, called a prism. A coloured band is then seen, in which the various tints are separated from each other (Fig. 20), and to this coloured band the name *spectrum* is given. In the case of the rainbow, we see the spectrum of the sun's light apparently projected in the air. The examination of various kinds of light with an instrument called a spectroscope (the essential parts of which are a prism, a slit to narrow the beam of light, lenses,

and an eye-piece) has shown (1) that all solid and
liquid bodies when heated sufficiently give out
all the various kinds of light, or, in other words,
that their spectra are all the same, whatever their
substance is, and are continuous bands of colour ;
and (2) that when gases or vapours are heated
sufficiently to give out light they only give out a.
few kinds, and that no two elementary substances
give out the same kind. In other words, the
spectra of glowing gases (*i.e.*, of the vapours of
metals, for example) are isolated bands of various
colours, in groups characteristic of each gas.
These facts are at the base of the whole science
of spectrum analysis, whether applied to the
detection of minute quantities of terrestrial substances on the earth, or to the recognition
of their presence in the atmospheres of the
sun, stars, comets, nebulæ, &c., since the mere
examination of any light enables the trained
observer to say with certainty whether the light-
source is a glowing solid or liquid on the one
hand, or glowing gas on the other, and if it be a
gas, to form an accurate notion of its nature. He
is also able to watch the constant changes going
on in the atmosphere of that source of nearly all
terrestrial energy, the sun,—to say approximately
the temperatures and pressures to which the
glowing gases are subjected, and, more wonderful
still, to estimate with tolerable accuracy the rate
at which some of the heavenly bodies are moving

towards and from the earth in the direct line of
sight !

Hitherto we have only considered the physical effects of heat upon matter, in which no
change takes place in the nature of the substance
itself, but only in its mood or condition. One of
the most important of the effects of heat, however, is that of promoting in matter those changes
which are known as chemical, in which more
than one kind of matter, or one substance, takes
part, and which result in the production of a
third substance different from either. For example, in every coal-cellar containing coal there
is in the fuel a large amount of potential energy
ultimately derived from the sun (Chap. VI.), and
the oxygen necessary for the combustion of that
fuel is present also; the fire, however, does not
burn. In order to bring into play the energy of
chemical attraction, the application of heat to
a portion of the coal is necessary, and when this
has once been done the chemical action continues,
a large amount of heat is produced, and at the
same time the coal disappears, being converted
into carbonic acid gas. In this conversion of
chemical energy into heat energy, which will be
specially considered in the next chapter, the heat
may be regarded as the mechanical result of the
collision of the atoms of the carbon and oxygen.

CHAPTER III.

THUS far we have been considering the effect of heat-energy upon one kind of matter or one substance at a time, changes in which the body undergoes no alteration in its kind, but only in its mood or condition. To changes of this nature the term physical is frequently given, in contrast to the term chemical, which implies the fact that two or more different kinds of matter are concerned in producing the effect observed, and result in the formation of a third substance differing in properties from either of the two with which the experiment is made. The present chapters deals with the relations between heat and chemical attraction, while the next one will deal, in part, with the relations between chemical attraction and electrical energy. We shall presently see evidence of the broad fact that whenever different substances combine under the influence of chemical attraction, heat is produced or evolved; and it will also be shown that, when it is desired to reverse this change, when it is wished to undo that work and to separate the two substances again, the application of heat to them

will in many instances effect that as well. In
short, we shall see how much "potential energy" is
stored up in substances between which there is a
strong chemical attraction, and how great an influence heat has upon the question whether these
different substances shall be attracted to, or repelled from, each other.

Although a strongly-marked distinction has
been drawn between physical and chemical
changes, it must be taken in the same sense, for
example, as when the difference between plants
and animals is exemplified by the instances of an
oak-tree and a cow. The tendency of all scientific
research is to obliterate these strongly-marked
lines of demarcation and classification, just as
among the lower forms of life it is frequently
a matter of doubt whether a given organism is
an animal or a vegetable (and, indeed, there is
the best reason to believe, in one instance at least,
that the same organism may be both animal and
vegetable at different periods of its life history),
so also are there actions, partly physical, partly
chemical, which it is difficult to assign to either
one class or the other.

Let us consider now a couple of simple illustrations of chemical action between two different
bodies, and the influence of heat upon them.
If a mixture be made of iron filings and flowers
of sulphur, a grey powder is produced, in which
the particles of iron and of sulphur each have

the properties that characterise larger masses of iron and sulphur; for example, a magnet applied to the mass will draw out of it all the iron filings. If, however, heat be applied to one part of it, enough to melt a very small portion of the sulphur, the whole mass will speedily glow with a bright red heat, and when it is cold it will be a dense compact mass, utterly unlike either iron or sulphur, and unacted upon by a magnet. If the experiment be made out of contact of air, the mass will be found to weigh exactly the same as the powder did before it was heated. Again, powdered charcoal and sulphur mixed together give another grey powder; if, however, the sulphur be (not melted this time but) turned into vapour in a closed vessel, and bits of hot charcoal be dropped into it, the charcoal and sulphur will unite chemically, and if the closed vessel be connected with a condenser, the result of the experiment will be found to be a clear bright liquid, as white as water, but very much heavier, known as carbon bisulphide, which is highly inflammable, and will dissolve many things which water will not.

In order, then, that chemical attraction may take place between the (atoms or) molecules of two substances, it is necessary to bring them into very close and intimate contact. This is very conveniently done by heat, which, as we have seen, tends to separate molecules from each other,

and thus to make easier the passage of two sets
of molecules between each other.* Any very fine
extension of their surface, or severe pressure,
however, will often bring two bodies within the
sphere of chemical attraction without any heat
whatever, a subject which has formed the basis
of some curious recent investigations. Instances
of some of these points will come before us in
this chapter.

Since we have now to deal with different
kinds of matter, it may be well to state here that
all kinds of matter known to the chemist are
either simple, *i.e.*, of one kind of substance only,
or compound, made up of two or more of the
simple ones. These simple substances, which
cannot by any known process be separated into
two others, are called elements. About seventy
elements are at present known, and of these
about sixty are the pure metals, iron, copper,
silver, &c. Of the remaining ten or twelve,
which are not metals, some are solid, such as
charcoal, sulphur, and iodine ; one is liquid (bro-
mine), while others are gases at the ordinary pressure and temperature, such as oxygen, nitrogen,
hydrogen, and chlorine. As examples of chemical
compounds we may cite chalk, made up of carbon,

* A rough illustration of this effect of heat was afforded by
the speaker presenting his right and left hands to each other, with
the fingers closed, when they could not interlace. Supposing heat
to expand the interval between the molecules—*i.e.*, the fingers—the
hands could then be interlocked.

oxygen, and the metal calcium; common salt, made up of sodium and chlorine; water, made up of oxygen and hydrogen; and bread, meat, and most foods, made up of the four simple elements—carbon, hydrogen, oxygen, and nitrogen. (The amount of energy to be got out of different kinds of food when we eat them, and the chemical changes which they undergo in our bodies, will be considered in Chap. VI.)

Having now a clear idea as to the difference between an element and a compound, let us consider some instances of the general statement that when two or more elements combine and produce a compound, their atoms rush together with great force, and the amount of heat developed by their collision is proportional to the mutual attraction of their respective atoms.

The most familiar example is presented by all cases of combustion, whether that be carried on merely for the domestic uses of warming and cooking, or for metallurgical purposes, as in the smelting of metals, or for the production of mechanical power by the aid of the steam-engine, or for the projection of missiles of war and the blasting of rocks, as in the combustion of gun powder. In all these instances, then, the heat-energy is developed from the chemical attraction between the fuel that is burnt and the element oxygen, which, in all cases except the last (to be

F

considered later), is supplied by the air around us, of which it forms one-fifth, and as a result compounds are produced which are neither fuel nor oxygen, but which contain both, and from which, by undoing the work done in combustion—by reversing the process, by unburning them, as it were—the oxygen and the constituents of the fuel may be recovered again.

Speaking broadly, all the substances that are used as fuel, or for the production of light (except electric lighting, which will be dealt with in Chap. V.) are made up of very little else than the two elements carbon and hydrogen, in different proportions. Charcoal, and that peculiar hard shiny coal called anthracite, are nearly pure carbon. Paraffin oil and coal-gas are almost entirely composed of carbon and hydrogen. Wood and bituminous coal, *i.e.*, the ordinary caking coal that we use, contain small quantities of oxygen, in addition to the carbon and hydrogen, as do also the animal and vegetable oils, and that elegant but expensive form of fuel, spirits of wine. The difference in properties of these various substances, and the various ways in which they behave when burnt, are largely due to the great differences in the relative proportions of hydrogen and carbon which they contain. Paraffin oil and spirits of wine, when burnt in similar lamps, behave very differently, the former giving a dull and very smoky flame, the latter a clear non-luminous blue one; and chemical analysis shows that the proportion by weight of carbon to hydrogen is as 4 to 1 in the first, and as 6 to 1 in the second. Smoke, as is well known, arises from the imperfect combustion of the fuel, due to a deficiency in the supply of air, or oxygen, and it is almost entirely composed of a mixture of other compounds of hydrogen and carbon, and of almost pure carbon. The formation of these products is due to the fact that in so many of our fire-places the coal frequently undergoes a sort of rough distillation (before it is actually burnt) without any attempt being made to condense the products. Before going farther into this question, however, it will be well to become acquainted with the chemical properties of oxygen, hydrogen, &c., and with the laws according to which these substances combine together.

One of the most important laws in chemistry is known as the law of combining proportions; it is the general expression of the fact that when two (or more) substances combine together to produce a third, they do so in certain definite quantities, or multiples of those quantities, which *never vary*. It is upon this invariability that the whole science of chemistry depends; but for it exact chemical analysis would be impossible. Each element has its own combining proportion,

F 2

and the numbers in the case of the three elements we are considering are—

Carbon	. 12
Oxygen .	. 16
Hydrogen . . .	1

Hence 12 parts of carbon combine with 16
parts (or a multiple of 16) of oxygen, and so
on. In this particular case the compound of
12 parts of carbon and 16 of oxygen is a colourless, poisonous, inflammable gas, called carbonic
oxide, which we often see burning with a pale
blue or yellowish flame on the top of a clear
coal fire; while the compound of 12 carbon to
(twice 16 or) 32 oxygen is the gas usually known
as carbonic acid, with which we shall shortly be
concerned. No other compounds of carbon and
oxygen only, are known, and should such be discovered they will be found to contain carbon and
oxygen, in the proportion of multiples of 12 and 16. .
The application of this law in the present
instance is that *when fuel is completely burnt*—

> 12 parts (by weight) of carbon take 32 parts oxygen,
> forming 44 parts carbonic acid.
>
> 2 parts (by weight) of hydrogen take 16 parts oxygen,
> forming 18 parts water.

Or, in this case by measure, oxygen being six-
teen times as heavy as hydrogen—

> 2 volumes of hydrogen take 1 volume of oxygen, form-
> ing 2 volumes of steam.

It is a very practical point, and one which cannot be too strongly or too clearly brought home to those in charge of furnaces, steam-boilers, &c., that (assuming for a moment that coal is nearly pure carbon, as it actually is in some instances) every 12 tons of coal require 32 tons of oxygen for their complete combustion, and as oxygen is only one-fifth of the air, they require 160 tons of air, or practically *a ton of coal re-quires 14 tons of air (or nearly 410,000 cubic feet) to be passed over it in order to burn it completely.* To move this mass of air requires the expenditure of considerable energy, and here we see one source of that loss of available energy pointed out in connection with the steam-engine.

Oxygen may be readily obtained in the pure state by heating some of its compounds in a closed vessel provided with an exit tube, the end of which dips under water, and the gas as it bubbles up (Fig. 21) is collected in a jar previously filled with water and inverted over the end of the pipe. Chlorate of potash, and oxide of manganese are the two compounds generally

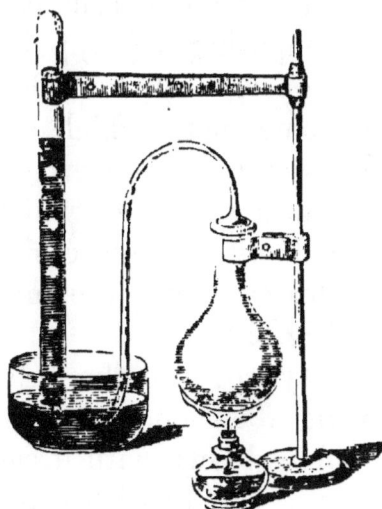

FIG. 21.

employed for this purpose.   Oxygen is the most
abundant of all the elements, forming eight-ninths
of the water, nearly one-fourth the air, and about
one-half of sand, chalk or limestone, and clay, the
three most abundant minerals on the earth's surface, as well as entering largely into the composition of most substances. Under ordinary conditions
it is a transparent colourless gas, but it has been
liquefied by cold and pressure.   It is not inflammable itself, but substances burn in it with much
greater brilliancy than in air, evolving a large
amount of heat energy.   A steel watch-spring,
for example, one end of which is heated white
hot in air, if plunged into a jar of oxygen, begins
to burn with brilliant scintillations.

Hydrogen is usually obtained from water by
decomposing it (i.e., pulling it asunder), either by
the energy of the electric current (p. 116) or
by some metal which unites with the oxygen of
the water, and turns out the hydrogen.   Zinc
put into water, to which a little acid has been
added, effects this easily ; heat is not required
in this case, and the mode of doing it is shown
in Fig. 22.

Hydrogen is the lightest substance in nature,
being about one-fourteenth as heavy as air ; soap-
bubbles blown with it easily ascend in the atmosphere.   It has also the greatest capacity for
heat, or specific heat (p. 38), of any known substance.   It is usually a transparent colourless gas,

but, like oxygen, it has been liquefied, and probably also solidified, by cold and pressure. Unlike oxygen, it will not allow substances to burn in it, but it is itself inflammable, burning in air with a pale blue flame, and producing water by its combination with oxygen.* The chemical attraction between hydrogen and oxygen is greater than between any other two known substances, and hence their combination produces the most powerful artificial heat which can be produced by purely chemical means. If the two gases be mixed together in the proportions by measure necessary to form water (p. 68), and a flame, or red-hot wire, be brought in contact with them, they unite with great explosive violence. In the oxy-hydrogen blow-pipe, the gases are brought to the jet by two separate tubes which unite, like a Λ, each provided with a cock for regulating the gas supply. The heat

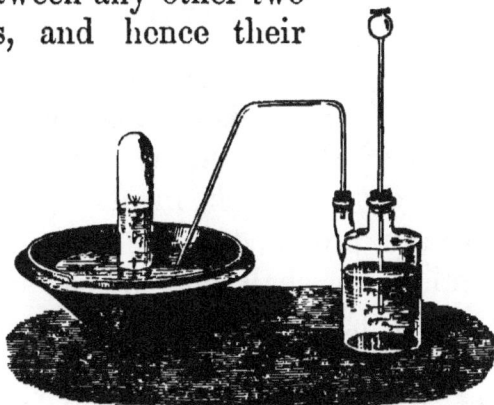

FIG. 22.

* The formation of water from the hydrogen of fuel or of gas when burnt, may be readily shown by holding a cold tumbler or bell-glass momentarily over a flame, and is also seen in the water which on a cold night trickles down the inside of a shop window where many lamps are burning.

of this jet melts some of the most infusible metals, and heats lime to intense whiteness, producing the lime-light. For this latter purpose coal-gas is frequently employed in place of pure hydrogen, with but little diminution in the effect.

The properties of the other element in fuel, carbon, in its usual form are familiar to most persons—a dense black solid, devoid of taste or smell.* When it unites with oxygen, the product is a heavy poisonous gas known as carbonic acid. This gas is exhaled naturally from various parts of the earth, as a result of the energies that are at work in its interior. It may be readily prepared in the gas-bottle (Fig. 22) by acting on chalk or limestone, which contain nearly half their weight of carbonic acid, by another acid, such as hydrochloric acid (spirits of salt). When limestone is "burnt" in a kiln, the heat expels the carbonic acid and leaves pure lime. Carbonic acid is also a transparent colourless gas, much more soluble in water than oxygen or hydrogen; it will neither burn itself nor allow anything else to burn in it, and it is poisonous to animal life. It forms the chief part of the fatal choke-damp or after-damp found in coal mines after an explosion. It is given off by all animals when breathing, and is one of the

* It occurs pure in nature, however, in a crystalline form, and is then known as the diamond. Jet, lampblack, charcoal, &c., as well as the diamond, are other forms of carbon, and they all produce carbonic acid when burnt.

results of fermentation and of the decay of animal and vegetable matter. A portion of Chap. VI. will be devoted to explaining the source of the physical energy of the human body, which is closely associated with the production therein of carbonic acid resulting from the union of the carbon in our food with the oxygen of the air.

The frequent formation and occurrence of this gas render it very desirable that a knowledge of its properties, and of some simple tests for its presence, should be widely spread. It is an exceedingly heavy gas, being about half as heavy again as air, with which it does not readily mix; this can be shown by actually pouring the gas from a vessel full of it into a vessel full of air, when the test of a burning taper will show that it has actually displaced the air, although there was no visible passage of matter from the first vessel to the second.* It cannot be too carefully remembered that air containing so much carbonic acid gas that a candle will not burn therein, is unfit also to support human life. A very simple test for it is afforded by clear lime-water (lime stirred up with water, and allowed to settle clear), which becomes milky directly it is shaken up in a bottle with air containing carbonic acid. The production of it in the

---

* One practical consequence of this is that it has a great tendency to remain at the bottom of old wells, mines, brewers' vats, &c.

human body may be shown by drawing air into
the lungs through lime-water, which remains
clear, and then breathing the same air out again
into the lime-water, which at once becomes
milky. The amount of carbonic acid in the
breath is about 5 per cent (five parts in 100),
and since pure air in the open country contains
only about three parts in 10,000, and in towns
seldom more than four or five in 10,000, while
air with only six parts of carbonic acid in 10,000
is felt at once to be close and disagreeable, the
need of the ventilation of confined spaces is very
evident. It has been shown by very careful experiments and calculations that in ordinary dwel-
ling-rooms of moderate size, the amount of fresh
air necessary to be passed through the room in
order to keep the proportion of carbonic acid
below 6 in 10,000 is about 3,000 cubic feet per
hour for each human being, for each lamp or
gas-burner, and for each pair of candles.*

It will now be desirable to consider a little
more in detail the conditions under which fuel is
burnt for the production of light, or, in other
words, the structure of flame. It may be stated
broadly at the outset that, under the ordinary
practical conditions of daily life, no flame is
luminous in which there are not some solid par-

* It is beyond the scope of this little book to go more fully into
this important question of respiration and ventilation, but the
reader will find it fully and yet popularly treated in Professor
Hartley's " Air and its Relations to Life " (Longman's).

ticles. The flame of burning hydrogen, for example, or of the mixture of coal-gas and air in a gauze burner or Bunsen lamp (Fig. 23), gives out scarcely any light; but if some iron-filings are sprinkled into it, or a metallic wire be held in it, these solid particles will be intensely heated, and will glow, or become incandescent, enough to give out light. Solids may be thus made to glow by the energy of the electric current, as will be seen when the "incandescence electric lamps" are explained (Chaps. IV. and V.). It has been shown above that all our fuel is essentially some compound or other of carbon and hydrogen, and also that the energy of chemical attraction between hydrogen and oxygen is much greater than that between carbon and oxygen. When, therefore, the combustion is so arranged that there is not enough oxygen present to burn both completely, the hydrogen is burnt first, and either pure carbon, or, as is more probable, certain other compounds of carbon and hydrogen containing a greater proportion of carbon than the original fuel, is left in the flame, to be raised to glowing point by the heat resulting from the collision of the atoms of hydrogen and oxygen. A careful study of a candle-flame (Fig. 24) will show that it consists in the main of three parts: (1) the exterior

Fig. 23.

shell, very faintly luminous, where the combustion of both carbon and hydrogen is complete; (2) the luminous part of the flame, containing the glowing and unburnt carbon compounds; and (3) the inner, non-luminous portion, where no oxygen penetrates, which consists chiefly of the gaseous fuel. If a cold plate, or even a piece of thick white paper, be suddenly depressed upon a flame, so as to cut it across the middle, a ring of black carbon (or hydrocarbon) will be deposited on it from the unburnt portions in the flame; and if the broken stem of a tobacco-pipe be inserted into the inner non-luminous part of the flame and steadily held in it (Fig. 24), inflammable vapours will issue from the other end, and can be ignited there. When a candle is blown out these vapours rise from the wick, and the candle can, with care, be rekindled by holding a light in these vapours several inches from it.*

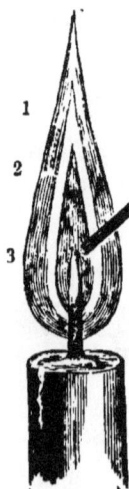

FIG. 24.

It appears, then, that the production of artificial light from fuel depends upon a proper

---

* These and many other interesting points are fully dealt with in Faraday's "Chemical History of a Candle" (Chatto and Windus).

adjustment of the supply of oxygen, *i.e.*, of air,
which varies with the nature of the fuel employed, and must be so arranged as to prevent
the escape into the air of unburnt particles of
fuel (*i.e.*, smoke), and yet to leave enough uncon-
sumed in the flame to give out light by their incandescence. The various arrangements of chim-
neys, globes, &c., around our lamps all have this
object in view. It is, moreover, a curious fact
that if the gaseous fuel be heated before it is
burnt, it produces much more light; this has
recently been taken advantage of by Sir W.
Siemens in the construction of large gas-lamps
of enormous power, rivalling the electric light.

We have seen that if hydrogen and oxygen
are mixed and heated to a given point, they
combine with explosive violence; the same thing
happens with a proportionately less development
of energy, when coal-gas or other inflammable
vapours are mixed with air, and more or less
heated. To this cause are due the lamentable
explosions in our coal mines; fire-damp, a mixture of gases, chiefly hydrocarbons, is pent up in
the coal, and is sometimes released, either by the
mechanical process of mining, or by a diminution
in the pressure of the air, owing to atmospheric
changes indicated by the barometer, a sudden
fall of which at certain periods of the year is
frequently followed in a few hours by a colliery
explosion. The fire-damp thus set free mixes

with the air, and when it comes in contact with a naked flame, or even with red-hot iron (heat, not actual combustion, being all that is necessary to kindle it), a very rapid explosive combustion takes place. Moreover, recent experiments have conclusively shown that the presence of finely-divided coal-dust in the air of the mine will communicate this rapid combustion, either alone or assisted by gas; and further, that *any* inflammable substance, if sufficiently finely-divided and suspended in the air, will communicate an explosive flame over a large area. Many fires in flour-mills, &c., have been traced to this cause. In fact, the rate at which a flame spreads is largely determined by the extent of surface available for contact between the combustible substance and the oxygen, and also by the heat-conducting power of the substance. Lead-foil, for example, cannot be made to burn in the air, but it is possible chemically to obtain lead in such a very fine state of division, that when air has access to it, it begins to burn of itself. The lead pyrophorus, or fire-bearer, is made by heating tartrate of lead in a glass tube till vapours cease to come off, and then sealing the tube. This action, due to increase of surface, is the explanation of most cases of spontaneous combustion. Oily rags and greasy sawdust, for example, present large surfaces of oil to the action of the air—they begin to absorb oxygen therefrom, the energy of chemi-

cal attraction is exerted, and presently the whole
mass bursts into flame. The same process goes
on in hay-ricks, and it is not the least of the re-
commendations of the new process of preserving
green fodder, called *ensilage*, that
the exclusion of oxygen is an
essential feature in it, fire-risks
being thus avoided.*

Allusion has been made to the
important influence of the heat-
conducting power of metals on the
spread of flame, and it is upon this
that the principle of the safety-
lamp (Fig. 25) used in coal mines
is based. Such lamps consist essentially of an ordinary oil lamp,
the flame of which is completely
surrounded with fine wire gauze,
through the meshes of which only
has air any access to the burning
oil. When the lighted lamp is
placed in an explosive mixture of

FIG. 25.

gas and air, the mixture is kindled and continues to burn inside the gauze cylinder, but
the metal wire conducts away the heat so rapidly
from the flame thus produced, that the gas on
the outside of the cylinder is not kindled, since

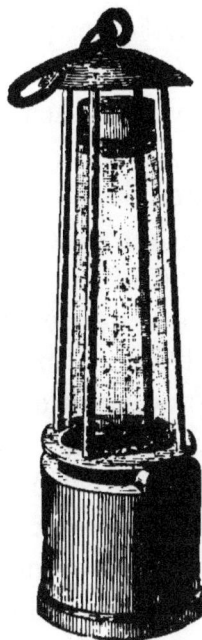

* The introduction of this process from America into this
country was recently advocated at the Society of Arts by Prof.
Thorold Rogers, M.P., in a paper which has since been extended and
published.

it is not heated to the necessary point of ignition.

From what has been said about the theory of combustion, it is evident that when it is desired to obtain the greatest possible amount of heat-energy out of fuel, that fuel must be *completely* burnt, so as to get the energy developed by the burning of the carbon as well as of the hydrogen. On a large scale this is far more practicable when the fuel is gaseous than when it is either solid or liquid, since the supply both of fuel and of air can be regulated to a nicety by valves, and hence it is that many furnaces now constructed for large metallurgical and other manufacturing operations are built in two or more parts, one of which is known as the gas-producer, the object of which is to roughly distil the coal, *i.e.*, turn it into gas before it is burnt. Great economy and many advantages result from this method. The cleanliness and convenience of coal-gas as a fuel for domestic use are now coming to be more generally recognised, owing to the perfection of the combustion causing no smoke, and to the absence of dust, &c., arising from the ash, or unburnt (and unburnable) mineral constituents of the coal; nor must the difference in the mode of delivery into houses of solid and gaseous fuel be overlooked. In the increased use of gas-stoves, gas-fires, and gas-cooking ranges, is unquestionably to be found the remedy for the smoke and smoke-fogs of our large towns, and proposals have been made by Sir W. Siemens and others to separate the gaseous products of the distillation of coal into two portions, collecting the first and last portions in one gasometer, and the middle portions in another, in order thus to supply lighting gas of higher illuminating power than at present (the middle portions), and, by a second set of pipes, gas of low illuminating, but great heating, power, for use in stoves, fires, &c.

In addition to the energy obtained *indirectly* through heat from chemical attraction, mechanical energy is often obtained *directly*, in the shape of an explosion. A simple case of this is the gas-engine (Fig. 26), which has been developed during the last few years, and marks an entirely new departure in the artificial development of mechanical energy. In these engines the piston is driven backwards and forwards in the cylinder by the explosion, either in the cylinder itself or in an adjoining chamber, of a mixture of coal-gas and air. The explosion in the early forms was very sudden and rapid, but latterly, by means of automatic valves regulating the supply of air and gas, it has been made more gradual, so as to produce a tolerably uniform motion. The chemical changes that occur in the action of this engine are somewhat complicated, and there is no doubt that the heat produced by the com-

G

bination, which expands the gases, largely increases the mechanical effect.

Directly connected with this subject is the
transmission of energy from one place to another
by the flow of gas. The use of the gas-engine

FIG. 26.

is probably at present the largest example of
transmitting power, and according to recent calculations power can be transmitted by gas at
one-twenty-fifth the cost of its transmission by
compressed air.

In most cases where explosive energy is exerted
for mechanical purposes, however, matters are so

arranged by the admixture of various solid or
liquid chemical substances, that under the influence of heat and of the chemical action induced
by heat, these various substances suddenly re-
arrange themselves in different combinations,
some of which in all cases are gaseous at the
temperature of their production, and it is to this
sudden production of gas, whose expansive force
is enormous, in a very confined space, that the
destructive action of explosives is due. Gun-
cotton, for example, cannot be distinguished by
the eye from ordinary cotton, but the influence
of a comparatively low degree of heat (whether
applied directly or derived from percussion) will
cause the atoms of which it is composed to change
their grouping instantly, and to form fresh compounds, all of which in this case are gaseous.
This is owing to the fact that, in making this
explosive, ordinary cotton (which contains carbon,
hydrogen, and oxygen) is soaked in nitric acid,
whereby it loses some of its hydrogen, and takes
up some nitrogen and a great deal of oxygen,
and this oxygen eventually helps to burn the
carbon and hydrogen. A similar change is made
in glycerine, during the preparation of nitro-glycerine, from which dynamite, litho-fracteur, &c.,
are manufactured by taking up the liquid explosive with some absorbent powder. In his annual
address as President of the Society of Chemical
Industry, Sir F. Abel stated that the manufacture

G 2

of dynamite had now (1883) reached the astonish-
ing total of 9,000 tons per annum. Blasting
gelatine, the " king of explosives," as it has been
termed by Sir F. Abel, is a solution of gun-cotton,
or nitro-cellulose, in nitro-glycerine. In the case
of gunpowder, the charcoal and sulphur which it
contains are burnt at the expense of the oxygen
in the nitre, a salt which contains nearly half its
weight of that gas. When powder explodes, the
gases produced by it, when measured at the standard temperature and pressure of 0° C. and 760
mm. (*i.e.*, 32° Fahr. and nearly 30in. barometer)
are 270 times the original volume of the powder,
but so much heat is produced by the chemical
action that if the explosion takes place in a space
confined to the original volume of the powder, a
pressure of 42 tons per square inch is exerted on
the walls of that space. The energy imparted to
a shot depends upon the rapidity with which the
powder disengages its gases, and slow burning
powder is, for artillery purposes, much better than
quick. If the energy were developed too rapidly,
it would be spent upon the powder-chamber,
instead of upon the propulsion of the shot, and
the gun would burst. The larger the "grain"
of the powder the more slowly does it burn, other
things being equal, and for very heavy guns the
"grain" is so large as to cause the term " pebble-
powder " to be employed. In the case of some
experiments in 1881 with the 100-ton gun, a

projectile of 2,000 lbs. weight was fired with a charge of 448 lbs. of pebble-powder, each pebble being about one inch in diameter. The striking energy of this shot as it left the muzzle was 33,500 foot-*tons*.

Hitherto we have been concerned chiefly with the production of heat when atoms collide, and chemical combination take place, although, in passing, instances have been mentioned where the decomposition or separation of substances was also effected by heat, as, for example, the " burning " of lime, when 50 parts of limestone lose 22 of carbonic acid, leaving 28 of lime. When a compound of two (or more) elements is resolved by great heat into its constituent elements, the process is called dissociation. Thus water may, if the steam be heated hot enough, be dissociated into hydrogen and oxygen; carbonic acid into carbonic oxide and oxygen, and so on. It is a remarkable fact that the other form of radiant energy which excites in our eyes the sensation of light, also effects both these kinds of chemical change. If a mixture of hydrogen gas and chlorine gas (which is best prepared by decomposing hydrochloric acid by electric energy, Chap. IV.) be exposed to light in a glass vessel, they will unite with explosive violence. On the other hand, a familiar example of the decomposition of compounds by light is found in the beautiful art of photography, where colourless substances containing silver are so affected by light that the image or picture produced by a lens is painted by the silver thus set free from its compounds. Another instance, which will be developed at some length in Chap. VI., is the decomposition, or dissociation, of carbonic acid by growing plants, under the influence of light, when the oxygen is set free, and the carbon enters into the substance of the plant, forming therein a store of potential energy.

CHAPTER IV.

THE first observation recorded with regard to this form of Energy in Nature was made by Thales of Miletus about 2,400 years ago, who observed that the curious substance amber, when rubbed, became temporarily endowed with the property

FIG. 27.

(which it had not previously) of attracting to itself light particles in its neighbourhood (Fig. 27). It is from the Greek name for amber ηλέκτρον, Electron, that the term Electricity takes its name. The next step was made by

Dr. Gilbert, in the reign of Queen Elizabeth, who, repeating the experiment of Thales, extended it to other bodies, such as diamond, rock-crystal, glass, sulphur, resin, &c., and showed that the same energy of attraction could be similarly excited in them, and he termed such bodies *electrics*. Otto Guericke, Sir Isaac Newton, and Dr. Franklin made further experiments; but many years elapsed before the various isolated observations were systematised, and brought into that condition of exact knowledge which we recognise as scientific. The science of electricity may be said to have been established by the labours of Franklin, Volta, Galvani, Davy, and above all, of Faraday, during the latter half of the last, and the earlier part of the present century.

It may be stated broadly that in every case of friction between (and probably even of contact of) two different bodies, there is a development of electricity. This is sometimes expressed in another way; it is said that " different bodies are at different potentials with regard to electricity;" the word "potential," in an electric sense, being used merely to express the degree in which a body is electrified. In the majority of instances, however, the effect is not perceptible, since the electricity passes away instantaneously by conduction, a process analogous to the conduction of heat (p. 40). Further

researches have shown that the phenomena observed are best explained (or, at any rate, best
described) by assuming the existence of two
kinds of electricity, of opposite properties, to
which the terms positive and negative, or vitreous
and resinous, are given;
the production of elec-
tricity by friction ap-
pears to be due to the
separation of the two
kinds of electricity, one

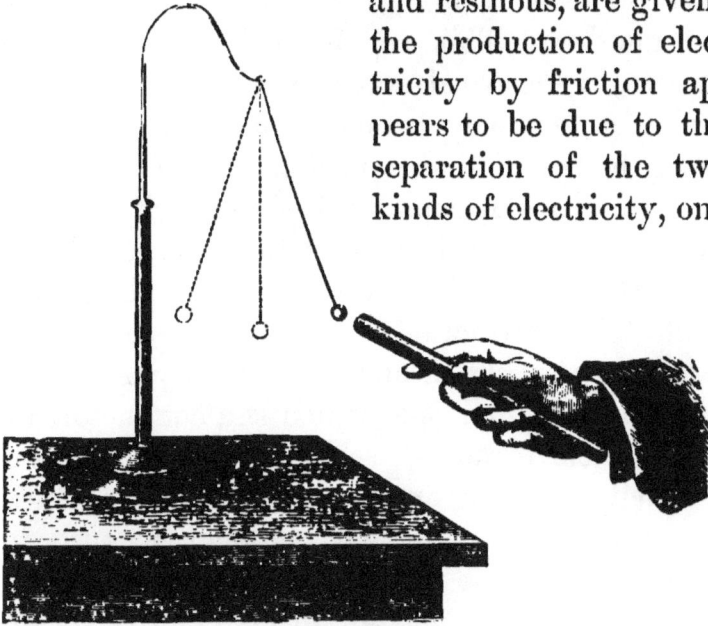

Fig. 28.

accumulating on the rubber, the other upon the
thing rubbed. If a pith-ball suspended by a
silk thread (which being a non-conductor pre-
vents the electric influence from passing away
from the ball) be touched with a piece of glass
that has been rubbed with silk, and the glass
be then withdrawn, the ball will no longer be
attracted by it, but repelled (Fig. 28). If then

a piece of shell-lac, resin, or sealing-wax, be rubbed with flannel and presented to the pith-ball, the ball will be attracted by it. From this we learn that bodies charged with the same kind of electricity repel one another, and that if charged with opposite kinds of electricity they attract one another. This repulsion is felt by many persons in electrical states of the air, their hair having a tendency to stand out from the head owing to the mutual repulsion between the fibres charged with the same kind of electricity. There are many mechanical operations, especially in the textile industries, in which these phenomena of electrical attraction and repulsion play an important part. When one excited body is brought near an unexcited one, which is a conductor, the first attracts towards itself the opposite kind of electricity existing in the second, and as these two opposite kinds have a tendency to rush together, they do so if the distance between them · be not too great, and an electric spark passes across the interval; the duration of this spark is not longer than the twenty-four-thousandth part of a second, and yet considerable heat, light, and noise are developed, and hence the energetic nature of electrified bodies is apparent. This apparent production of an opposite electrical state in bodies brought near to a substance previously electrified is known as induction.

When it is desired to obtain larger supplies of

electric energy than are produced by rubbing rods of glass, shell-lac, &c., electrical machines are employed, all of which consist of two parts, one for producing, the other for collecting, the electricity (Fig. 29). The producing part consists usually of one or more

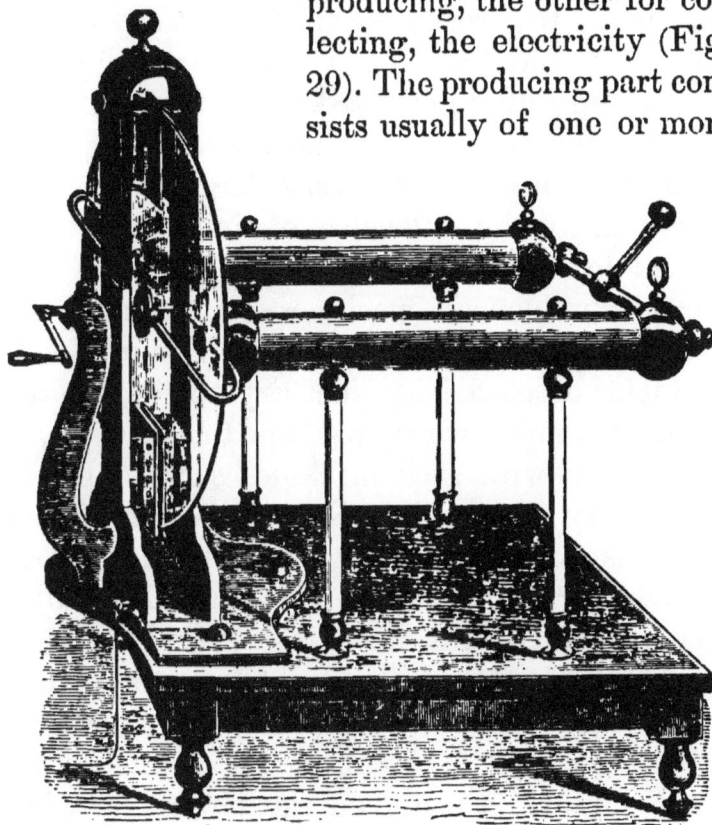

Fig. 29.

plates of glass, to which a rapid movement of rotation can be given, during which it is rubbed; the collecting part comprises metal points and conductors, placed near the rotating plate, and mounted upon some non-conductor, as a stem of glass or ebo-

nite. A description of the details of their construction and mode of action would occupy more space
than is here available, but it may be found in any
elementary treatise on electricity.* They may be
regarded generally as contrivances to develop electrical energy of a peculiar kind at the expense of
mechanical energy. The most recent types of
these machines depend very largely for their
effects upon the influence of induction above
alluded to, and are known as "induction" and
also as "influence" machines. In consequence
of the fact that damp air is a very much better
conductor of electricity than dry air, or, in other
words, is a much worse insulator than dry air, all
experiments with such machines succeed best in
a warm dry room. The electrical effect of these
machines may be stored or accumulated in a Leyden jar (Fig. 30), which is simply a wide-mouthed
glass jar coated with tinfoil inside and outside for
about three-fourths of its height from the bottom,
and provided with a wooden cover, through
which a metal rod passes, having a knob on its
outer end, and a chain on the lower end which
lies on the inner tinfoil coating. When a charge
of one kind of electricity is driven into the jar,
by connecting the rod with an electrical machine,

* "Elementary Lessons in Electricity and Magnetism," by
Silvanus P. Thomson (Macmillan, 1882), Ferguson's "Electricity"
(Chambers, 1882), Balfour Stewart's "Primer of Physics" (Macmillan), and Wormell's "Electricity and Magnetism" (Murray,
1882), may be consulted with advantage for further information
on the subjects of this and the following chapter.

it spreads over the inner surface, and drives the whole of the other kind away on to the outer surface, whence it passes into the earth in contact with it. The jar may be retained in this condition for some hours, or even days; but whenever a path is provided by which the two kinds

of electricity can re-unite, they will flow to one another, and the jar will be discharged. If this

FIG. 30.

path is through the human body a shock will be felt; if it is through a metal rod and across a space of air (Fig. 30) the discharge will be sudden, and accompanied by sound, light, and heat (from the transformation of electrical energy); if it is through a fine wire or a wet string, the discharge will be slower and more

quiet.   It is important to remember that, as was
first shown by Dr. Franklin, who constructed a
jar with movable coatings, the charges of the
jar really reside in the glass itself, and not in
the metallic coatings. These effects may be
largely increased by employing a number of
Leyden jars in a " battery," when energetic
mechanical and heating effects may be accomplished by the discharge, such as the perfora-
tion of cards, and of plates of glass, the fusion
of metallic-foil and wire, the combination of
mixed gases, the ignition of combustibles, explosives, &c. By far the grandest exhibition of
these, however, is seen in the lightning-flash,
the destructive energy of which is well known,
and with good reason feared.   This flash, which
is a discharge either between a cloud and the
earth, or between two clouds, at the point where
the air offers the least resistance, may sometimes be a mile in length, and its duration is not
really more than the 1–100,000th part of a second,
although the impression on the retina of the
eye lasts much longer.   Its electrical energy is
sometimes changed into mechanical, as when
buildings are destroyed ; and sometimes into
heat, as when metallic wires and rods in its path
are melted.   Dr. Franklin, who first established
the identity of atmospheric electricity and that
of the machine, suggested in 1749 the use of
pointed metallic rods to protect property from

destruction by lightning—thus affording a path by which the discharge should take place quietly. Care should be taken in fixing these rods that their tips are of some metal which does not tarnish in air, that the points project well above the highest part of the building, that metal work about the roofs be connected with them by stout wires, and, most of all, that their lower ends lead into damp ground ; the neglect of this last precaution, which is frequently unattended to, will make the best-laid conductor *practically useless*, while the owner sleeps in fancied security. A rough practical rule for the height of a conductor is, that it will protect a circular area at its base whose radius is equal to its own height, so that a rod 50 feet high will protect a circle on the ground round it whose diameter is 100 feet.

Electrical energy may be developed in many other ways than by friction; a violent blow, and even steady pressure, produces opposite electrical states on the two opposing surfaces—the tearing of paper or linen, the crushing of sugar, the cleaving of a sheet of mica, all produce it. Many bodies in passing from the liquid to the solid state become electrical, the phenomena of combustion and evaporation are attended by it, and in the evaporation of water over the surface of the oceans is seen one source of atmospheric electricity. Certain crystals (*e.g.*, tourmaline) when heated are found to develop

opposite electrical charges at opposite poles.
Many animals (notably the electric eel), and
some plants, produce electrification, and Volta
showed that the mere contact of certain metals
caused them to assume opposite electrical states.
Hence, as has been pointed out by Fleeming
Jenkin,* " a sense enabling us to perceive electricity would frequently disclose a scene as varied
as a gorgeous sunset . . . Every movement
of our body, each touch of our hand, and the
very friction of our clothes would cause a play of
effects analogous to those of light and shadow on
the eye. . . . Without eyes we might never
have discovered the existence of light. By direct
perception we have become aware of the vast
importance of light, and it is probably owing to
the absence of direct perception that we do not
yet know the part which electricity plays in the
economy of nature."

Thus far we have been considering chiefly
the production of opposite electrical states in
bodies, of static electricity, *i.e.*, of electricity at
rest; the only instance of electricity in motion
being afforded by the spark, or discharge, passing
between two oppositely electrified substances, as
in the Leyden jar or the lightning-flash. It will
be desirable now to study a little more in detail
some of the other modes of producing electricity

* S.P.C.K. Manuals of Elementary Science: "Electricity"
(pp. 51-53).

in motion, *i.e.*, electric currents, since they afford
remarkable instances of the general doctrine of
the conservation of energy.' The transformation
of mechanical into electrical energy will form the
chief subject of the next chapter, the remainder
of the present one being devoted to the connection between heat and moving electricity, and
between chemical attraction and moving electricity.

It has been already pointed out that the mere
contact of two different metals gives rise to
opposite electrical states in them, but so long as
there is no difference in temperature between
various parts of their junction there is no discharge, or movement of electricity—no current is
produced. If, however, heat be applied to the
point of contact of any two dissimilar metals, and
their free ends be united by a wire, a current of
electricity will be found to flow through the wire
and through the point of junction, in a direction
varying with the pair of metals employed. This
phenomenon is known as thermo-electricity, and
it was first observed in 1822 by Seebeck. The
best mode of detecting the existence of the current is by its action upon a magnetic needle (to
be fully explained in Chap. V.), which is turned
to one side or the other of its normal position
when an electric current circulates near to it.
An instrument for subjecting a magnetic needle
to the influence of a current is called a galvano-

H

meter, and such an arrangement is shown in
Fig. 31. The intensity of the effect depends
upon the metals employed, and upon the temperature; the pair that produce the greatest
effect are bis-
muth and anti-
mony. If a
series of these
bars be solder-
ed together in
such a way
that all the
odd-numbered
joints are on
one side and
all the even-
numbered on
the other, a
thermo-elec-
tric battery is
formed, and
the electrical

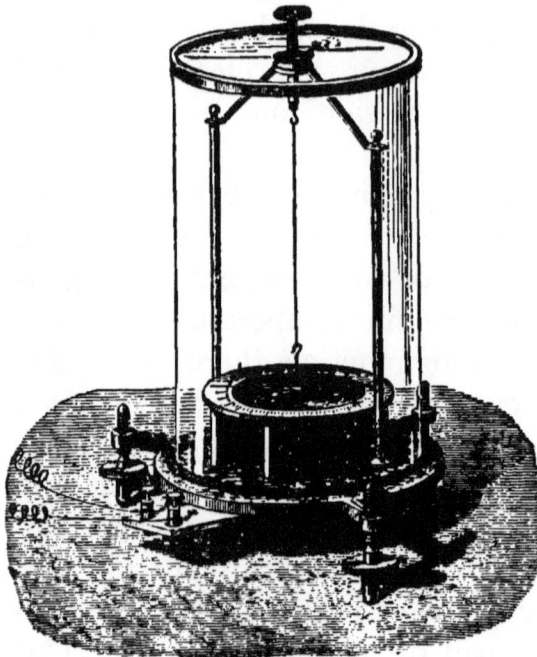

FIG. 31.

energy of the arrangement depends upon the difference in temperature which can be maintained
between the two sets of joints. Such batteries
have been constructed in a form powerful enough
to produce the electric light, and other familiar
effects of strong currents. When made on a very
small scale the arrangement is known as a thermo-
electric pile (Fig. 32), and in combination with a

delicate galvanometer is an exceedingly sensitive instrument for detecting minute changes in temperature, being largely used in researches upon radiant heat. It should be noted here, as another

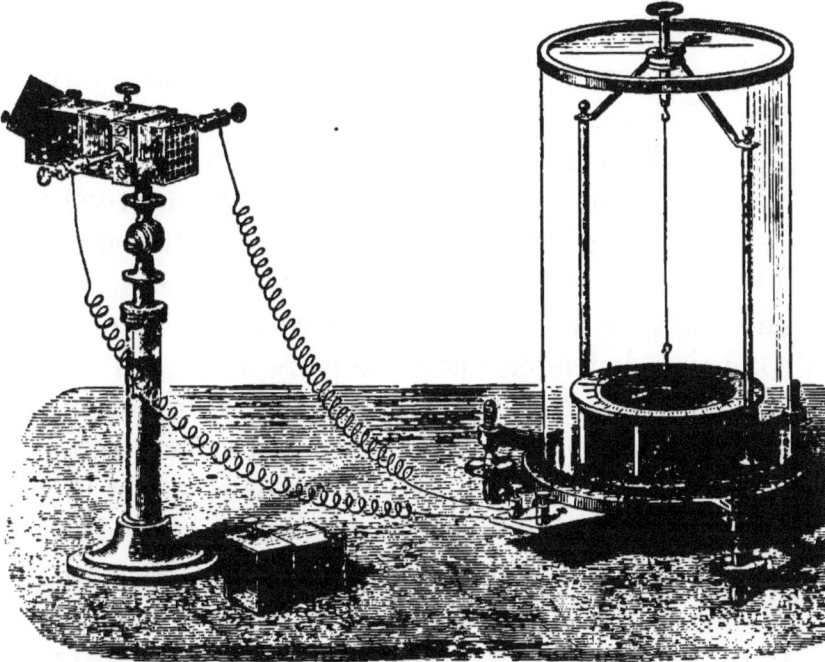

Fig. 32.

illustration of the conservation of energy, that when an electric current is passed through a junction of dissimilar metals, the junction is either heated or cooled, according to the direction of the current.

We have now to consider the connection between chemical attraction and electricity, or, as it is often called, the production of electricity

H 2

by chemical action. This was due, in the first instance, to two Italian men of science, Galvani and Volta, who early in the present century investigated this subject, which we now recognise as one of the transmutations of energy, and whose names are perpetuated in the terms Voltaic and Galvanic batteries. Volta showed that when any two dissimilar metals are brought into contact, each of them is found to be in an opposite electrical state to the other, one becoming positively (+), and the other negatively (—), electrified. The amount of difference between these states, and whether any given metal was in a + or — condition depended upon the pairs of metals employed—iron, for example, being positive towards copper, silver, and gold, and negative towards tin, lead, and zinc. The reality of the existence of this "contact force," as it is often called, was doubted for a long time, but such doubts have recently been set at rest by some very delicate experiments of Sir W. Thomson's. It will be noticed, however, that this effect is a mere *state*, or condition; in order to produce a *flow*, continuous discharge, or *current* of electricity, something more is necessary, and this is found in the arrangement known as the voltaic cell (Fig. 33). To construct it take a vessel of water, to which a few drops of sulphuric acid, or a few crystals of common salt or of sal-ammoniac, may be added with advantage, and place therein two plates, z, c, of dissimilar metals,

taking care to keep them apart. A very usual
and effective pair is copper and zinc, with sulphuric acid and water. As long as the two plates
are kept apart scarcely any effect is observable,
and none if the zinc be quite pure or be previously
rubbed over with mercury, *i.e.*, amalgamated. If,
however, a wire, M, be attached to each plate,
and examined as to its electrical
condition, the two wires are
found to be in opposite electrical
states, and as soon as they are
joined the two electricities *rush
together, and continue to flow or
circulate* along the wire. In the
case of zinc and copper, the
wire attached to the copper is
called the positive pole, but,
as in Volta's contact experiments, the intensity of the effect

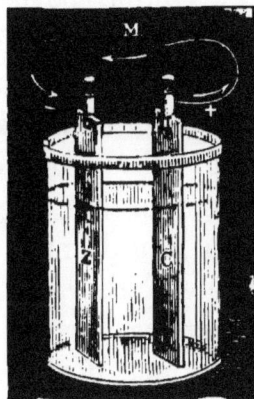

Fig. 33.

produced, and whether any given metal is + or —,
depends on the particular pair of metals employed.

It will presently be seen that a wire joining
two plates under these conditions, is in a very
different state or mood from an ordinary wire.
It does not weigh any more while in this state,
but it has many curious properties—chemical,
magnetic, and physiological—and these are expressed by saying that a current of electricity
circulates or flows in the wire. It is, however,
important to remember that this is merely a

convenient phrase to express a set of facts. *We do not know that anything actually flows* along the wire, although there are some reasons for believing that these observed effects are due to a peculiar condition of vibration, or motion, set up in the wire, different from those accompanying the manifestations of heat-energy.

One remarkable fact, however, in connection with this production of electrical energy by a voltaic cell, is invariably noticed, viz., that it is *always accompanied by chemical action* going on in the cell. One of the two metals must have a considerable chemical attraction for oxygen, and the liquid must be one capable of acting on the metal; there is, however, no proof that their electrical behaviour is due to their chemical behaviour, nor *vice versâ ;* but the two sets of phenomena invariably occur together. When zinc foil is burnt in the air, or a mass of zinc in a crucible is heated by fire, the zinc is oxidised at the expense of the oxygen of the air, and heat energy is produced. When, however, zinc is oxidised at the expense of the oxygen of water (and hydrogen is given off), which is the chemical change that occurs in the voltaic cell, electrical energy is produced.

Volta increased these effects by arranging a number of the cells in a series, in the manner indicated in Fig. 34, to which the name of " crown of cups " was first given, and this is the

principle of the arrangement of several cells of the same kind in a Voltaic battery. It was soon found that the bubbles of hydrogen gas evolved stuck to the plates (especially to the copper one, from the surface of which they chiefly rose, although produced at that of the zinc plate), and that the energy of the cells rapidly became less. Numerous plans have been devised to overcome this difficulty, usually by the adoption of two fluids, separated by a

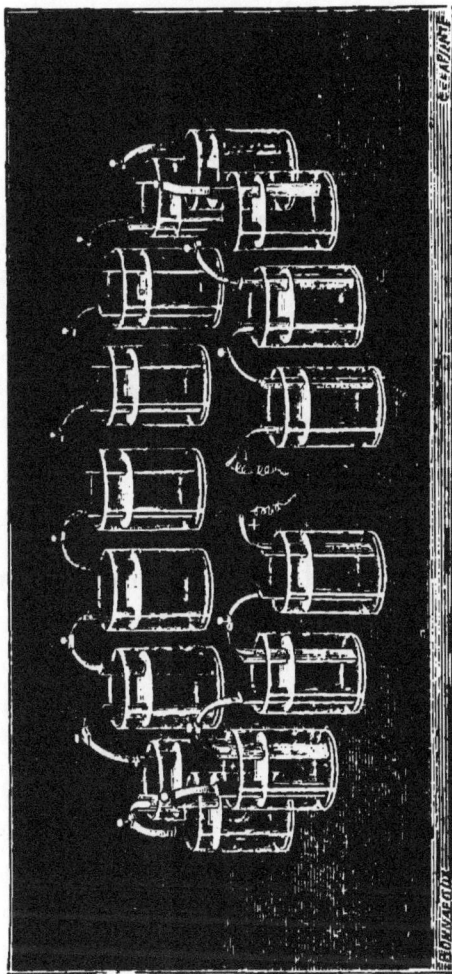

Fig. 34.

diaphragm of porous earthenware, as well as two metals. Daniell's constant battery (Fig. 35) has the copper-plate immersed in a solution of copper sulphate (blue vitriol), the zinc plate and dilute

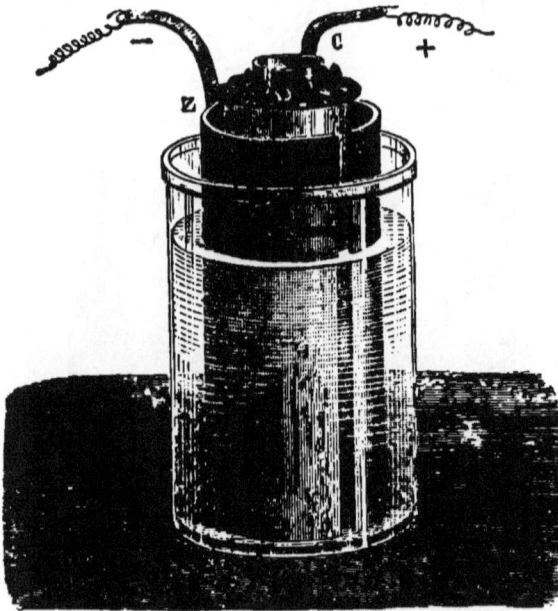

FIG. 35.

sulphuric acid being contained in a porous vessel, through which the hydrogen passes, and instead of coming off as gas, deposits copper on the copper-plate, keeping its surface always clean and bright. Nearly 88,000 cells of this type are employed in the British postal telegraph service. In the Leclanché battery (Fig. 36) the only exciting liquid is a solution of sal-ammoniac, and the porous vessel, M M, contains a carbon rod, C, surrounded by a mixture of carbon and oxide of manganese : 56,000 of these are used in the same service. In the so-called bichromate cell, the plates are zinc and gas-

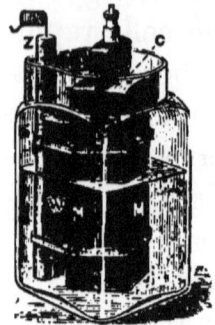

FIG. 36.

carbon, surrounded by a mixture of bichromate of
potash and sulphuric acid (Fig. 37). More than
20,000 of a modification of these are used in the
post-office telegraphs. Many other forms have
been devised, the most energetic for a short period
being Grove's nitric acid battery, in which the
two metals are zinc and platinum, and the two
liquids are weak sulphuric acid
and strong nitric acid. Each
arrangement has advantages of
its own, and is best suited for
particular kinds of work.

It will be convenient here to
define certain terms that are
frequently employed in connection with electrical energy.
In a single cell or battery, the
path of the positive current
from the zinc through the liquid,
copper-plate, and wire back to
the zinc again, is spoken of as

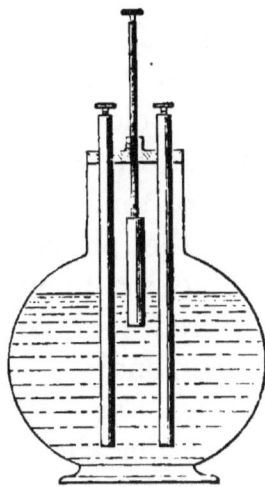

FIG. 37.

the electric circuit, consisting of the liquid part and
the metallic part. When the metallic circuit is not
continuous, it is said to be broken, and unless the
current has enormous energy it will not leap over
the smallest break of continuity; when two surfaces
touch so closely that the current passes from one to
the other, they are said to be in contact. A conductor is a substance along which the current flows
more or less freely, such as most metals; and it

is a curious fact that those metals which conduct
heat best (p. 40) also conduct electricity best :
the order of conducting power is the same for
both.   An insulator is a substance through which
the current will not pass, such as silk, glass,
earthenware, gutta-percha, &c., and insulators are
used for preventing the electric energy from leak-
ing out of the conductors.   The phrase electro-
motive force (for brevity often written E.M.F.)
denotes that which moves, or tends to move
(p. 4) electricity from one place to another,
just as the pressure in a system of water-pipes
sets, or tends to set, the water in motion along it ;
and the phrase " potential" is used to express
the degree to which a body is electrified, a great
difference of potential between any two bodies
corresponding to considerable E.M.F. between
them.   For further information on these subjects,
the reader should consult the various text-books
on Electricity before referred to.

It will be well now to consider some of the
effects of electricity in motion, or current electricity, and to observe how they differ from those
of electricity at rest, or the static electricity of
bodies in opposite electrical conditions.   In the
discharge of the Leyden jar, we noticed some of
the work that was done by moving electricity,
but the discharge and its immediate consequences were of excessively short duration.   In
the case of current electricity, however, the

effects produced by it last as long as the current continues to flow, or, in the case now being considered, as long as the chemical changes go on in the cells of the battery, provided also that the circuit is maintained untouched, and that contact is nowhere broken.

It should be carefully borne in mind, however, that the phrase " current of electricity " is purely a conventional one, for there is no proof that anything " flows " along the wire. Our actual knowledge is confined to the fact that a wire under these conditions possesses certain remarkable properties, and that this change, whatever it be, is communicated along the wire at a speed closely approaching the velocity at which light travels, considerably exceeding 150,000 miles in one second.

We have seen what occurred when resistance was offered to the energy of mechanical motion—viz., that it was converted into heat-energy. The same thing happens when resistance is opposed to electricity in motion. This resistance is offered by bad and small conductors, and accordingly we find that when a current of electricity meets a high resistance in its path, the place where that occurs is more or less heated. The experiment may be effectively made by introducing into a circuit along which a strong current is flowing, a short fine wire, too small to convey the whole of the current, when it will be seen that the wire will get intensely

hot, and if the energy of the current is sufficient
it will melt, and the circuit will be broken. This
power of exciting heat-energy at will by means
of electricity is, as is well known, extensively
used for firing mines at a safe distance, thus
avoiding the possible accidents from time-fuses ;
for discharging heavy artillery, either singly or
simultaneously, in a broadside on a man-of-war,
and, as will shortly be seen, it is at the basis of
all applications of electricity to lighting purposes,
and more especially to that form known as incandescence lighting. The resistance offered by a
conductor depends upon its size, and also upon
the material of which it is made. A very pretty
experiment illustrative of this latter point is to
construct a chain of alternate links of the same
gauge and lengths of silver and of platinum wire.
When a strong current is sent through this, the
platinum links become red-hot, while the silver
links, being better conductors—*i.e.*, offering less
resistance—remain comparatively cool.

The principles of electric-lighting will be
explained in the next chapter, since, as practically carried out at present, they involve those
questions of the relations between magnetism
and electricity which will there be specially considered. It will be sufficient to remark here
that there are, broadly, two systems—[1], the
Arc system, in which the light is produced when
the resistance opposed by two pieces of carbon

(with a thin stratum of air between them) is introduced into the circuit; and [2], the Incandescence, or glow, system, in which the necessary resistance is given by a continuous thin filament of carbon interposed in the circuit. It may be remarked here that neither system is as new as is generally supposed, the arc-light having been produced by Sir H. Davy three-quarters of a century ago,* and an incandescent carbon lamp having been publicly exhibited in Birmingham by Mr. W. Mattieu Williams more than thirty years ago; in all such cases, however, the electrical energy was developed by chemical means, which were so costly as to prohibit the use of the light except for scientific experiments, and on occasions when expense was no object. The reason why so much has been heard during the last few years of the application of electricity to lighting, and to various other purposes of practical life (such as motive power, &c.) is, that only recently have the means been discovered of transforming the cheapest source of energy known to us, viz., mechanical, into electrical, and also of effecting the reverse change. The apparatus which effects this is called a dynamo-machine, and will be fully explained in the next chapter. It may be interesting here to note the

* The first production of an arc-light is probably due to Etienne Gaspard Robertson, and is recorded in the *Journal de Paris*, for the date 22 Ventôse, An X (March 12, 1802).—Vide *Nature*, June 7, 1883

comparative cost of electrical energy developed
chemically and mechanically; the following table
gives the number of foot-pounds of energy which
can be got out of one ounce weight of different
substances :—

Gunpowder	100,000
Coal	695,000
Zinc	113,000
Copper	69,000
Hydrogen	2,925,000

From this it appears that coal is capable of
giving out six times as much energy as zinc,
so that even if coal and zinc were the same
price per ton, the (electrical) energy produced
by zinc would be six times as costly as the
(mechanical) energy produced by coal. As is
well known, the price of a ton of zinc is many
times that of a ton of coal, and hence, even
under the most favourable circumstances, the
cost of electrical energy developed chemically
compares most unfavourably with the cost of
that developed mechanically, in the manner to
be described in the next chapter.

We have seen that the mere approach of an
electrified body towards a conductor in a neutral
condition promotes, or induces, an opposite electrical state in that other body; a phenomenon of
a similar kind in current electricity was observed
by Faraday in 1831, and the fundamental fact
of current induction may be thus stated. If two

wires are laid parallel to, but insulated from, each other, and a current be sent along one of them, it is found that at the very instant when the current commences to flow in it, a momentary current passes along, or is induced in, the second wire. This induced current is in the opposite direction to the current in the first wire, and it ceases immediately, no change occurring in the second wire until the current in the first wire ceases to flow; at this instant another induced current makes its appearance in it, but in the reverse direction to the first induced current, and therefore in the same direction as that of the continuous current in the first wire. The same effects are observed when a wire along which a continuous current is flowing is brought near to, and removed away from, another separate wire. The experiment is best shown by insulating both wires with silk or gutta percha, and winding them in coils on thin wooden bobbins (Fig. 38); the ends of the first, or "primary" wire, are connected to a battery, and the ends of the second to a testing instrument called a galvanometer (Chap. V.). When the primary coil is put inside the secondary, and when it is removed from it, evidence is obtained of the development of the induced currents above described. The same effect may be produced in a more perfect way by allowing the primary coil to remain inside the secondary, and fitting to the former a little mechanical arrangement

which makes and breaks contact rapidly with
the battery, and consequently produces a very
rapid succession of these induced currents. The
primary coil is a short length of comparatively
thick wire, of low resistance, and the secondary
coil is composed of many turns of very fine
and well-insulated wire. Such an arrangement

Fig. 38.

is known as an Induction coil, and sometimes
as a Ruhmkorff's coil, from the name of a very
celebrated maker of them (Fig. 39). Induced
currents always possess great electro-motive force,
and their sparks will strike across intervals that
no battery will reach; their physiological effects
are very strong, even comparatively small medical coils giving most unpleasant shocks, while
the discharge from some large coils is sufficient
to kill a man. The largest coil yet constructed

belonged to the late Mr. Spottiswoode, President
of the Royal Society (1883); it is 4 feet long
and 20 inches in diameter, weighing altogether
15 cwt.; its primary coil is 660 yards of wire
nearly $\frac{1}{10}$th inch in diameter, and the secondary
coil contains 280 *miles* of wire, nearly $\frac{1}{100}$th inch

Fig. 39.

in diameter; when excited by a Grove's battery
of 30 cells it gives a spark $42\frac{1}{2}$ inches long, a
veritable miniature flash of lightning!

Many of the effects of the ordinary "electrical
machines" (p. 91) can be produced by these coils,
but their chief use is for the study of the electric
discharge under various conditions either in air
or in a partial vacuum, some of the experiments
upon which last are probably the most beautiful
in the whole range of experimental physics. The

I

character of the luminous discharge is subject to
almost innumerable variations, depending upon
the degree of exhaustion of the glass tube or
vessel (Figs. 40, 41), and
upon the kind of gas or
gases contained in it at these
exceedingly minute pres-
sures, but it is a remarkable
fact that when the exhaustion has reached the stage
of a nearly perfect vacuum,
the discharge will not take
place at all. The study of
these discharges has thrown
considerable light upon that
wonderful natural phenomenon, the Aurora Borealis,
and also upon the molecular
theory of matter referred to
in Chap. I. The very re-
cently published investigations of Prof. Selim Lein-
ström (vide *Nature*, Nos.
709, 710, &c.) into auroral

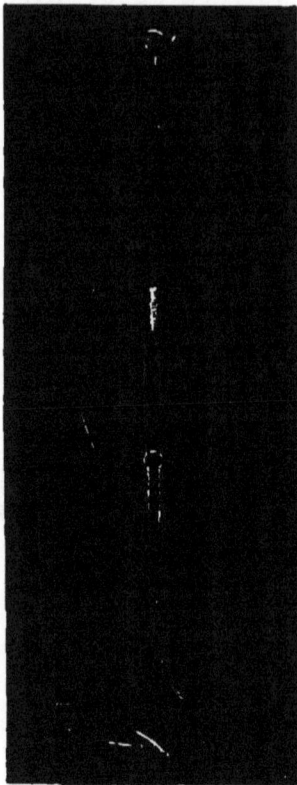

Fig. 40.

phenomena, have thrown considerable light upon
that subject.

In the last chapter we saw several instances
of the heat-energy developed by the force of
chemical attraction, and it was also pointed out
that under certain conditions the process could

be reversed, *i.e.*, that if heat-energy were allowed
to act on the compounds thus produced, the work
could be undone, and the compounds separated
or decomposed. Now it is found that electrical
energy is one of the most powerful agents with
which we are acquainted for decomposing
chemical compounds, especially when they are

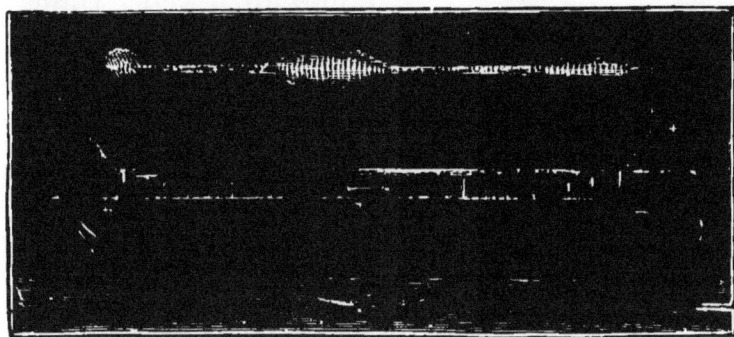

Fig. 41.

in the liquid state. No liquid (except melted
metal) which conducts electricity at all, conducts
it without being thus decomposed, and when
this occurs the electrical energy is spent in
overcoming the resistance of the chemical attraction for each other of the two substances
forming the compound. The most familiar example of this is the electrical analysis, or the
electrolysis, as it was called by Faraday, of
water. *Pure* water belongs, like turpentine,
petroleum, and many oils, to the class of non-
conductors, but if a few drops of sulphuric acid
be added, it may readily be decomposed by the

I 2

current from a few cells of a battery. The
arrangement is shown in Fig. 42; the wires from
the battery are attached to two platinum plates,
over each of which is inverted a tube closed at
one end, and filled with water. When the con-
nections are made, bubbles of gas rise from each
plate to the top of the tube, displacing the water,
and in a short time it
will be evident that
twice as much gas is

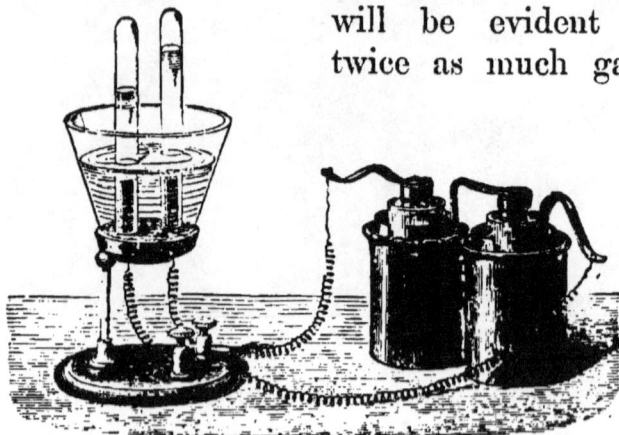

Fig. 42.

collected in the tube which is connected with
the zinc end of the battery, as in that connected with the copper end. If the tubes be
then removed and their contents examined by
chemical tests, the larger volume of gas will be
found to be hydrogen, and the smaller, oxygen,
thus demonstrating the statement of the composition of water given on p. 68. If, instead of
water alone, a solution of any metallic salt be
similarly treated, as, for example, sulphate of

copper (blue vitriol), sulphate of nickel, acetate of lead (sugar of lead), &c., oxygen will still be given off in one tube, but in the place where hydrogen before appeared no gas will be given off; instead of it the platinum plate will be covered with a deposit of the metal itself, copper, nickel, or lead, according to the salt employed. These facts form the foundation upon which the whole art of electrotyping and electro-plating has been reared, to which industry the phrase electro-metallurgy is often given. The articles to be plated are connected with the zinc end of the battery, or with the corresponding end of some other source of electricity, and are suspended in a tank containing a solution of the metal to be deposited on them. In the case of silver-plating, a double cyanide of silver and potassium is used, and for gilding, a similar salt of gold in a hot solution. The other end of the battery (or dynamo-machine, Chap. V.) is attached, not to a platinum plate, but to a sheet of the metal whose salt is in solution, which is also hung in the tank. When the circuit is completed, the electrical energy tears asunder the metal from the acid in the salt, depositing it upon the substance to be plated, and the acid and oxygen thus set free at once exercise their attraction for the metallic plate, which is gradually dissolved away, and thus the strength of the solution is kept up. Electrotyping is the name given to the art of copying seals, medals, engraved plates, &c., in copper, and may be readily practised by the amateur. A sharp mould must first be taken of the object to be copied, in fusible metal, sealing-wax, gutta-percha, or plaster of paris, and if it be made of any except the first named, it must be rubbed over with plumbago (black-lead) to make the surface conduct electricity. The mould is then hung in a tank as above described, containing a solution of sulphate of copper and a sheet of copper, connected with a battery, and left until a sufficient thickness of metal has been deposited, when the electrotype and the mould are pulled asunder. A large number of the cuts in this book have been printed from electrotypes thus produced. For small objects a simpler arrangement suffices, known as the single cell apparatus (Fig. 43). It consists of an earthenware vessel half filled with a strong solution of sulphate of copper, in which also stands a porous earthenware pot containing dilute sulphuric acid and a plate or rod of zinc, which is connected by a wire to the mould hung in the solution, and the strength of this is kept up by

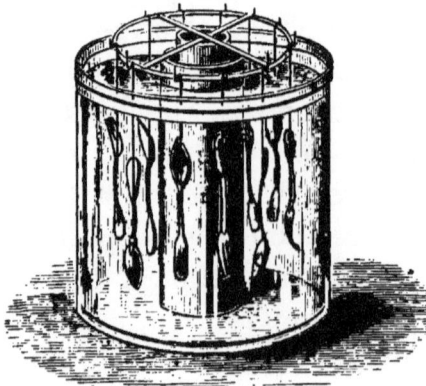

Fig. 43.

suspending some crystals of sulphate of copper in it.

We have seen that whenever chemical combination takes place some form of energy is developed, and also that energy is absorbed in undoing that work. Now, since the amount of energy produced in the first of these processes varies with the quantity and with the kind of the materials employed, but is always the same when the same quantity of similar materials are used, we should naturally expect to find that the amount of electric energy necessary to undo this work varied in the same way, and this is actually found to be the case. In other words, the *work done* is proportional to the electricity generated, and to the amount of zinc "burnt" in the battery. When water is thus decomposed, the amount of gas evolved depends upon the strength of the current employed, and upon the time for which it acts, and Faraday constructed an instrument called a Voltameter, in which the mixed gases are measured in a graduated tube, and from the quantity collected in a given time, the strength of the current can be calculated. Mr. Edison has recently proposed to apply the same principle to a meter intended to measure the amount of electricity drawn from the town-supply by any private house, whether for lighting or as motive-power (Chap. V.). In this ingenious instrument the weight of copper deposited by a known

fraction of the current upon a plate of metal
gives the necessary data for the calculation, and
by a curious mechanical contrivance the instrument is provided with a set of dials, to be read
off just like a gas-meter. But further, if the
same strength of current, *i.e.*, the same quantity
of electric energy produced by the consumption
of the same amount of zinc, is used to decompose
salts of various metals, it is found that the amount
of metal deposited differs in each case, and, in
fact, that these different weights stand to each
other in the same relation as the numbers which
express (pp. 67, 68) the combining proportion or
equivalent of each element. Thus, for every $32\frac{1}{2}$
parts of zinc consumed in the battery (or deposited by a given strength of current) $31\frac{1}{2}$ parts
of copper, 108 parts of silver, $29\frac{1}{2}$ parts of nickel,
and $65\frac{1}{2}$ parts of gold would be deposited. Moreover, as hydrogen develops the greatest amount
of energy in its combinations, it is not surprising
to find that the greatest amount of energy is
necessary to decompose its compounds; accordingly only one part (by weight) of hydrogen is
produced by the consumption of $32\frac{1}{2}$ parts of
zinc.

Recently, the electro-chemical decomposition
of water, or rather of weak sulphuric acid, has
assumed exceptional importance in connection
with the so-called "storage of electricity," or
"storage of force," erroneous expressions, as will

be seen in the sequel, for which should be substituted the phrase " electrical storage of energy." The first observation on the subject is due to Sir W. Grove, who, in 1842, noticed that if he decomposed water in a voltameter, and then, having disconnected the battery, joined the two platinum plates by a wire (Fig. 44), a current of electricity passed along the wire, and at the same time some of the gas disappeared. Moreover, the current was from the oxygen to the hydrogen, or in the reverse direction to that which separated them. A battery made up of fifty of these cells possessed sufficient energy to decompose water, and even to produce the electric light, and their

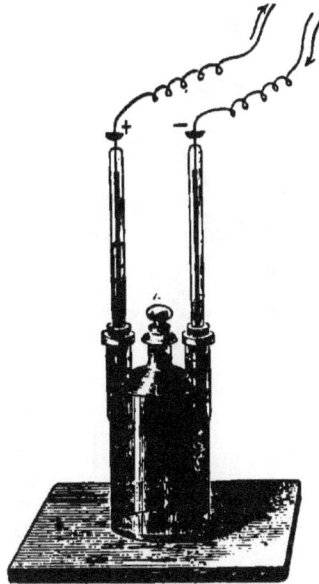

Fig. 44.

energy appears to be due to, at any rate it is always *accompanied by*, chemical action between the hydrogen and oxygen, just as in the simple voltaic cell, chemical action between the zinc and oxygen (p. 102) always accompanies the production of current electricity there.

The first person to apply this fact upon an industrial scale was M. Gaston Planté, who in 1860 constructed his storage-battery, or accu-

mulator, in the following simple manner:—Two
long wide slips of lead are laid upon each other,
separated by narrow strips of gutta-percha; the
whole is then rolled up in a spiral form and immersed in a (glass) jar containing weak sulphuric
acid; a connecting strip is at-
tached to each plate, and carried outside the cell (Fig. 45).
When the current from a battery (or a dynamo-machine) is
sent through the cell, water is
decomposed, and the plate by
which the current enters (*i.e.*,
that attached to the " copper,"
or positive, end of the battery)
is covered with a thin film of
oxide of lead, while the hydrogen is absorbed by the other
lead plate. When the battery·
is disconnected the cell may
be kept some time in the same
condition, but as soon as the
two lead-plates are joined by a wire a current
passes between them, and at the same time some
of the hydrogen combines with the oxygen of the
oxide of lead on the other plate. If the cell be
now charged again, but in the reverse direction,
more gas will be absorbed by each plate, and if
this process of charging and discharging alternately in opposite directions be repeated several

Fig. 45.

times, the surface layers of both lead-plates will get into such a porous condition that they will absorb very large quantities of gas. This process is known as "forming" the cell, and is somewhat tedious. To obviate this M. Camille Faure in 1880 coated both plates with red-lead, and thus obtained the necessary porous condition of metallic lead with a much less number of charges and discharges of the cell than was necessary in Planté's form. A plan of the arrangement of the plates in a Faure cell is given in Fig. 46. In the Faure-Sellon-Volckmar accumulators—probably the most efficient yet (June, 1883)—produced on a commercial scale, thick leaden gratings are cast, the perforations in

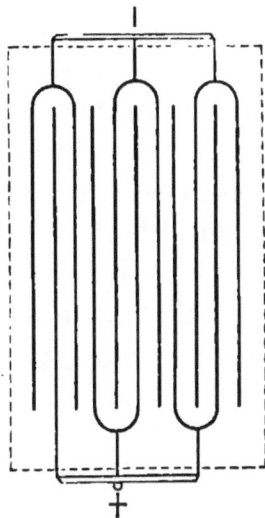

Fig. 46.

which are then filled with oxides of lead; and, indeed, nearly all the modifications proposed have for their object the increase of surface of the lead employed, upon a support which is not acted on by the liquid contents of the cell.

Opinions are somewhat divided as to the exact nature of the chemical changes that occur in these cells, but there can be no question that

sulphuric acid, sulphate of lead, peroxide of lead
and water play an important part therein, and that
they are continually undergoing *de*composition
and *re*-composition.   The main points, however,
to be remembered in connection with them
are—that when the cell is being charged the
electric energy does chemical work in the cell,
and in so doing ceases to be electricity ; that
the work which is done is the overcoming of
the chemical attraction of certain substances
for each other ; that the cell in the charged
condition resembles an ordinary voltaic cell,
from which electric energy can be drawn at
pleasure by completing the circuit outside it ;
and that when the cell is discharged, and the
electric energy is drawn off, the electricity is
produced at the time, and is always accompanied
by, and probably caused by, the re-union of
the substances previously separated.   Hence .
these accumulators *do not contain electricity* as
such, but only the means of producing it when
wanted—energy, not electricity, is stored in
them.   A precisely parallel case to this will be
found in Chap. VI., where the storage of solar
energy in wood and coal (through the medium
of growing plants), and its reproduction as heat
and light when the carbon and oxygen thus
separated are brought together again, are described at some length.   In fact, in almost all
cases a store of energy arises from the separa-

tion of two bodies which desire to come together.

Nearly all the practical uses of these accumulators, or secondary batteries, are dependent
upon the employment of dynamo-machines for
charging them, the electric energy being thus
produced cheaply. Their use in this connection will be alluded to in the next chapter, but
it may be convenient to state here the amount
of energy that can be thus stored electro-chemi-
cally. The numerical relations between mechanical and electrical energy will be explained in
the next chapter, and hence, without giving the
intermediate steps, it may suffice to say that a
fully-charged Planté cell can store about 11,000
foot-pounds of energy per pound of lead in the
cell, and that a Faure cell can store nearly
13,000 foot-pounds in the same weight of lead.
According to Sir W. Thomson a single Faure
cell of the spiral form, weighing 165 lbs., can store
2,000,000 foot-pounds of energy. Prof. Henry
Morton, of the Stevens Institute of Technology,
New York, in a report (dated February 17,
1883) upon the Faure-Sellon-Volckmar accumulators, states that this same quantity of 2,000,000
foot-pounds of energy, or one horse-power working for one hour, can be stored in a cell of
that kind which weighs only 80 lbs., and that
three cells, while standing fully charged for six-
teen days, only lost by leakage 7 per cent. of

the energy contained in them. The latter part
of this result the author can corroborate from
his own experience of the working of these cells,
under very trying conditions.

CHAPTER V.

BEFORE proceeding to the study of the very intimate relations existing between these two forms of Energy in Nature—relations upon which all the applications of electricity to the purposes of practical life depend—it will be well to consider briefly some of the elementary phenomena of, and some of the terms used in connection with, Magnetism. The force of attraction in this form of energy was first noticed many hundred years ago in certain hard black stones found in Magnesia in Asia Minor (whence the name *magnet*), but afterwards in other parts of the world. This lodestone (*i.e.*, leading-stone) as it was called, was afterwards shown to be a peculiar iron ore, a combination of iron and oxygen, and although it was at first thought that its attractive influence was confined to iron and steel, further experiments (as in the case of electric attraction) showed that a large number of other substances were thus affected by it, though in a less degree. In the tenth or twelfth century it was noticed that such stones when hung up

by a thread pointed in definite directions, North
and South, rapidly taking up that position again
when disturbed from it, and shortly afterwards
it was found that iron and steel, when rubbed
in certain directions with a piece of lodestone,
acquired these same properties. In 1600 Dr.
Gilbert published the results of very careful experiments with magnets, adding greatly to the
knowledge then existing. It is, however, perhaps scarcely necessary to say that even at the
present day we are as ignorant of the nature
of magnetism as we are of electricity—none of
these forms of energy are recognisable apart
from matter, as has been already pointed out.
There are strong reasons for believing that the
phenomena of magnetism are in some way connected with the motion of the particles of those
bodies, which, like iron, become magnetic; that,
in fact, it is another form of the molecular motion
spoken of in Chaps. I. and II. On this view of
the case, which has quite recently received strong
support from experiments exhibited to the Royal
Society by Prof. D. E. Hughes, the difference
between the arrangement of the particles in a
magnet and in an ordinary piece of steel or
iron, might be likened to the difference in the
packing arrangements of two boxes of eggs—
in the first (corresponding to the magnet) the
eggs are carefully packed, lying side by side
parallel to each other and to the sides of the

box, with their small ends all turned in the same
direction, and therefore touching the larger end
of the adjoining egg; while in the second (ordinary iron or steel), badly packed, the separate
eggs lie in all sorts of positions with regard to
each other, and at all angles of inclination to
the sides of the box. Another and quite different
physical theory of magnetism will be alluded
to later; but it may be noted with advantage
in this connection that the magnetism of iron
and steel is always materially lessened, and sometimes entirely destroyed, by changing the molecular condition of the iron; this may be done
by subjecting a magnetic rod to a mechanical
twist, or strain of any kind, or by heating it,
all magnetism disappearing at a cherry-red heat.
The influence of temperature upon magnetism
is well seen in the case of manganese, which
only becomes magnetic when cooled to below
zero Fahr.

If a bar-magnet, *i.e.*, a straight piece of steel
which has been magnetised, be carefully examined, by holding the different parts of it near
to a small nail suspended by a long thread (Fig.
47), it will be found that its attractive force is
most strongly exerted by the two ends, to which
the term poles is given. Moreover, if one pole
be thrust into a heap of nails, and withdrawn
again, it will be observed that not only do **many**
nails adhere to the magnet itself, but that other

J

nails adhere to them, and that quite a string of
them may be thus drawn up; if, however, the
nails actually in contact with the magnet be
removed from it, the attraction of the other nails
for each other at once ceases.　This shows that a

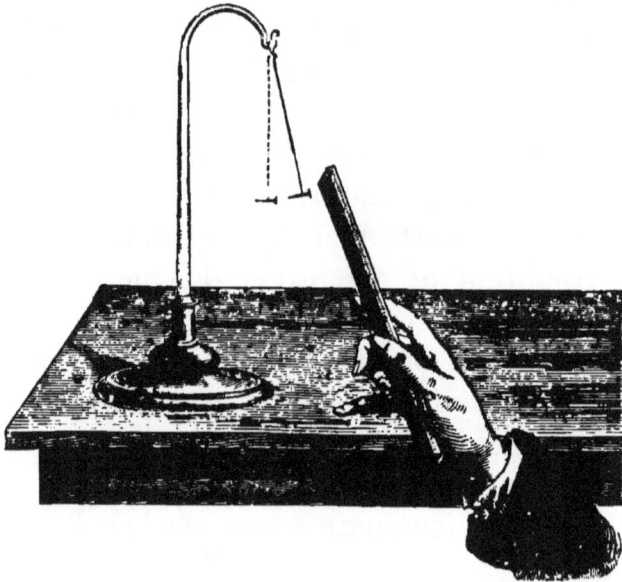

Fig. 47.

magnet is capable of inducing magnetism temporarily in other pieces of iron somewhat in the
same manner as is observed in the case of electrified bodies (p. 90).　Further, this attractive
force between a magnet and iron is mutual, for if
a magnet which is free to move be brought near
to a fixed piece of iron, the latter will attract the
magnet (Fig. 48).

It has been already stated that a bar-magnet, freely suspended by the middle of its length (either by a loop of string or on a point of support), will take up a definite position. If, now, another bar-magnet be brought near to one end (say that

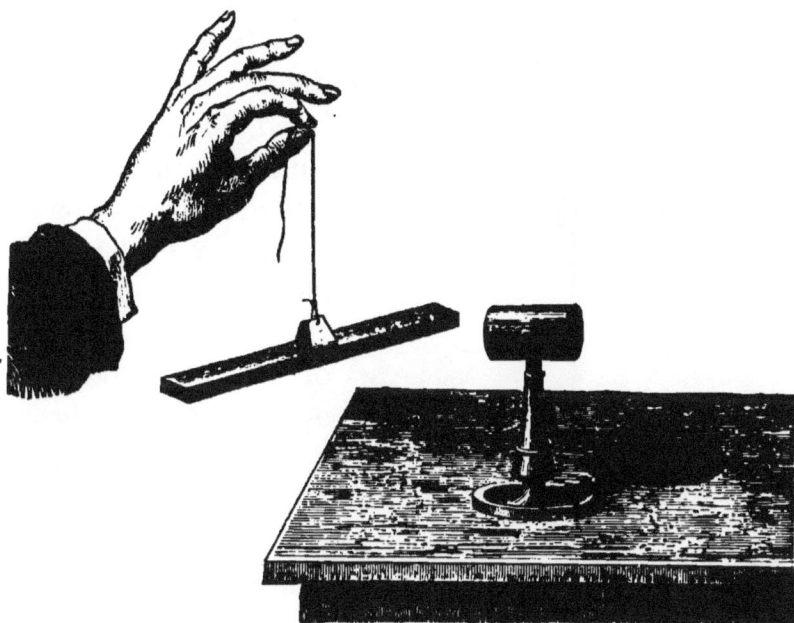

Fig. 48.

pointing northwards) of it, the two ends of the second magnet will behave very differently towards it, and careful observation will show that when two poles of the same kind (*i.e.*, two **N**-pointing or two **S**-pointing) are brought near to each other, they tend to repel each other, while between two poles of opposite kinds there is

j 2

attraction. These phenomena are of the utmost possible consequence in the theory of the dynamo machine, and in the transformation of mechanical work into electricity, and *vice versâ*, and when the attractions and repulsions are strong the energy brought into play is very great indeed. On the other hand, no matter how small the pieces into which a magnet may be broken, each piece has opposite properties at opposite ends, or is said to be endowed with polarity. In these phenomena of attraction and repulsion, again, the careful reader will note the analogy with corresponding electrical conditions.

FIG. 49.

The north-pointing property of a magnetic needle (Fig. 49) has been known and used for the purposes of navigation for more than 2,000 years. The mariner's compass consists essentially of a magnetic needle balanced on a pivot and supporting a card on which are marked the "points," and the whole is enclosed in a box

which is mounted on two rings with pivots called
gimbals, in such a way as to be independent of
the motion of the ship. The use of the compass,
however, is not nearly as simple as might be
supposed. In the first place, it is affected by the
iron of which most ships are built, and by that frequently contained in the cargo, and in the second
place, careful and exact observations show that
the direction in which the needle points is itself
subject to constant change. It varies slightly
with the time of day; it varies a good deal with
the particular place on the earth's surface where
it is observed, and it varies from year to year in
the same place. These different changes, as well
as the fundamental phenomena of **N.** and **S.**
pointing, are best explained by regarding the
whole earth as a huge magnet, whose magnetism is constantly varying in different parts
of it.

The subject of terrestrial magnetism is a
large one, and is being daily investigated at ob-
servatories, but space does not admit of its further
discussion here at present. It is, however, interesting to note here that the regular changes in
the earth's magnetism are associated in some way
with those periodic changes in the atmosphere of
the sun which we recognise as sun-spots, and also
with displays of the aurora borealis, and other
phenomena.

It will now be convenient to examine the

directions in which the lines of force emanate
from a magnet, and these are best seen in the
study of what Faraday called magnetic fields.
A magnetic field is simply that portion of space
which is within the sphere of action of a magnet.

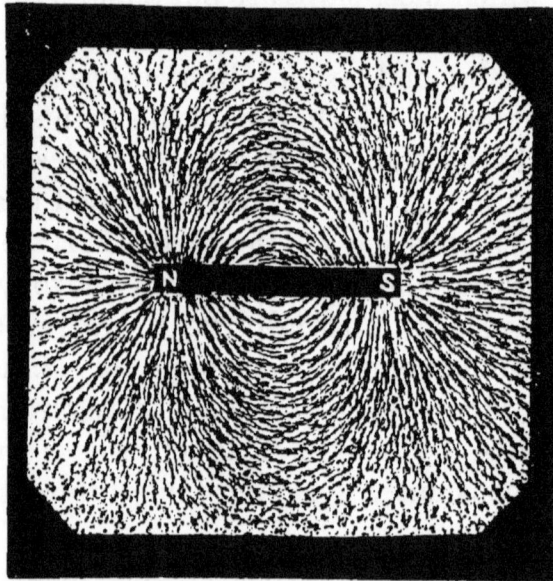

FIG. 50.—MAGNETIC FIELD OF A BAR-MAGNET.
(*By permission from "Engineering."*)

If a card or piece of stiff paper be laid upon a
magnet, and very fine iron filings be sifted there-
upon, the iron filings will (especially if the card
be gently tapped) take up certain definite positions
(Figs. 50, 51) indicating the lines of magnetic
force in the field.

We now approach the connection between
magnetism and electricity, which may be thus

stated broadly :—*A wire conveying electricity acts
like a magnet.* In Fig. 52 we see the " magnetic
whirls " surrounding a conducting wire—the
magnetic field of that wire, in fact. One
direct consequence of this, that a copper wire,

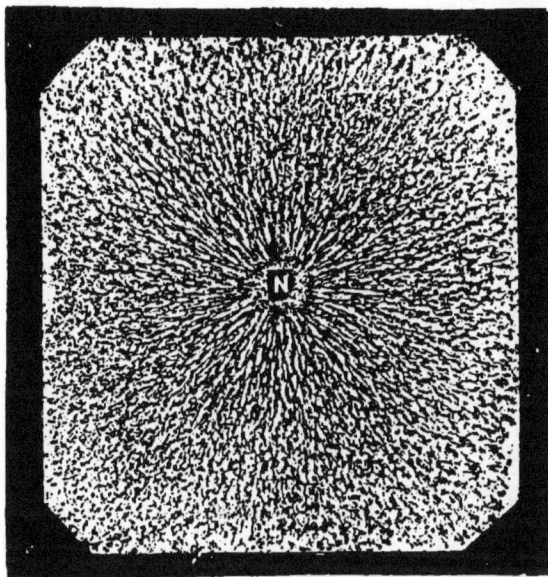

Fig. 51.—MAGNETIC FIELD OF ONE POLE OF A BAR-MAGNET PLACED
" END-ON."
(*By permission from* " *Engineering.*")

along which an electric current is flowing, will
attract iron filings to itself, is very easily
shown by making the experiment in the same
way as with a magnet. Moreover, if a conducting wire be delicately balanced on a pivot
(Fig. 53) like a magnetic needle, it will be found
to be attracted and repelled by the poles of a
magnet, according to the direction of the current

in it; and further, if two such wires be presented to each other, all the experiments of attraction and repulsion performed with two magnets, can be repeated with them. Fig. 54 shows the in-

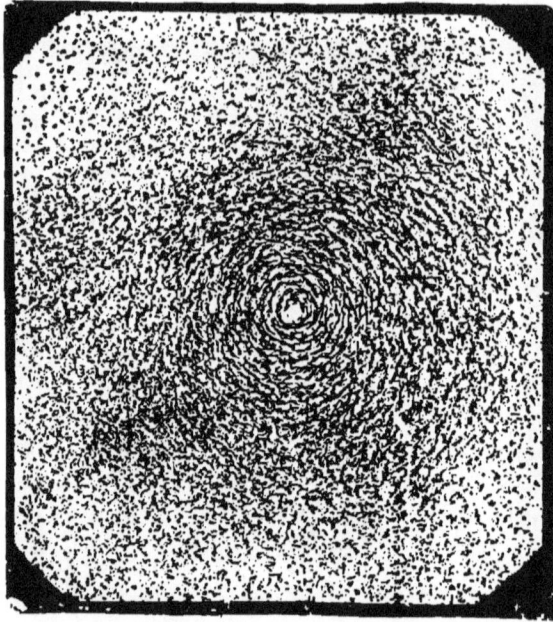

fluence which the direction of the current has on the experiment; parallel currents passing in the *same* direction *attract*, and passing in the *contrary* direction *repel*, each other. Lastly, if a long hollow spiral coil of insulated wire (technically called a solenoid) be suspended so that it is free to move, when a current is sent along it, it will behave like a magnetic needle, and take up a

N. and **S.** position. On these facts Ampère founded a theory of magnetism, according to which electric currents were perpetually circulating in the same direction round the particles of the magnetised body (Fig. 55), and to these currents the magnetism was due.

The discovery of the influence of a conducting wire upon a magnetic needle was made by Oerstedt in 1820, who found that if a stout wire were laid parallel to such a needle mounted on a pivot, as soon as a current passed along the wire the needle tended to place itself across it (Fig. 56). It should be noted

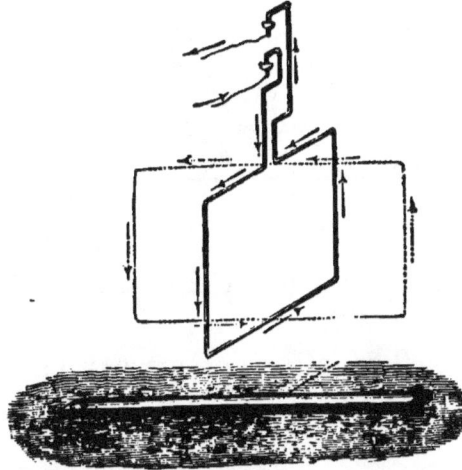

Fig. 53.

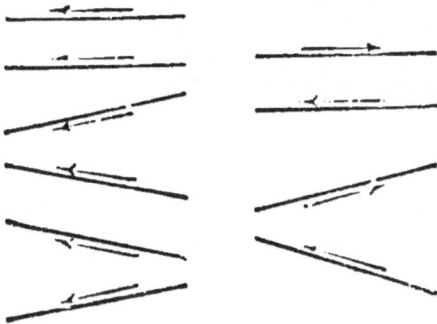

Fig. 54.

that this was the first instance of mechanical

motion being produced by current electricity. The side to which the needle turns, or is deflected, depends upon the direction in which the current flows in the wire, and upon its position with regard to the **N.** and **S.** pointing poles of the magnet. It may be determined by Ampère's rule, which is fully explained in the text-books referred to on p. 92.

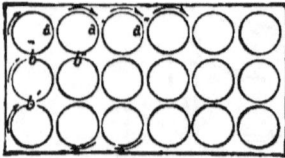

FIG. 55.

The effect of a weak current may be largely

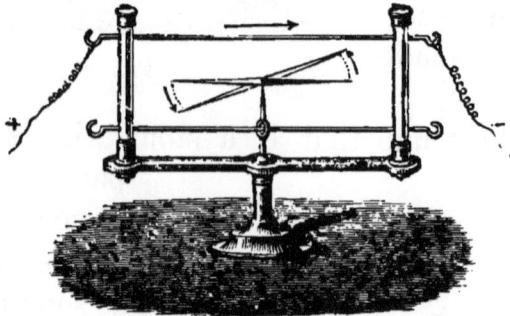

FIG. 56.

increased by winding insulated wire several times round the needle, thus causing the current to pass over it in one direction, and underneath it in the opposite (Fig. 57), and still greater sensitiveness is given by mounting a pair of parallel needles on one axis, with their poles, *a*, *b*, and *a,' b'* (Fig. 58), in opposite directions. If the needles are equal in magnetic strength, they are said to be *astatic*, and the combination being uninfluenced by the earth's magnetism, has no tendency to point in any particular direction, and is therefore much more sensitive to electric currents.

On this principle— the action of a current upon a magnetic needle—are constructed galvanometers (Fig. 59), instruments to detect the presence of, and in some modifications to measure the force of, electrical currents. On the same principle also depends one form of the electric telegraph, that known as the single-needle instrument, in which letters are indicated by beats to the right and left of the needle on the dial underneath which are the coils of wire. These beats are produced by sending the current from the distant station in two different directions, by a mechanical arrangement of springs operated with a handle. This was the earliest practical instrument devised, and it is still largely employed,

FIG. 57.

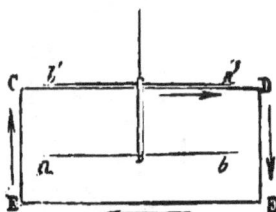
FIG. 58.

both by the British Post-office and on railway lines, for which it is peculiarly adapted,
about 16,000 being in use for this latter purpose
alone.*

No single letter requires more than four beats

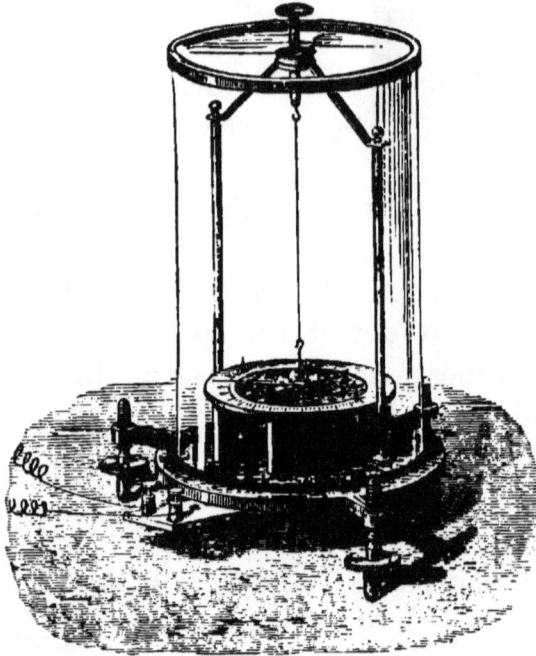

Fig. 59.

of the needle, and an expert operator can transmit
from thirty to thirty-five words per minute in
this way.

* For a popular account of Electric Telegraphy, consult "Telegraphy," by Messrs. Preece and Sivewright, in Longmans and Co.'s
series of Text-books of Science.

Having seen the influence of the magnetic field of a conducting wire upon a magnet, let us now examine its influence upon soft iron. The fact that a magnet induces magnetism in soft iron in contact with it, leads us to expect that current electricity will do the same thing, and experiment justifies that expectation. Soon after Oerstedt's discovery described above, Sturgeon showed that soft iron placed in a coil of wire through which a current was passing, induced magnetism in itself from that "field," but that this was only temporary, and that as soon as the current ceased, the iron lost its magnetic properties.

Steel thus treated, however, is permanently magnetised, though in a less degree; in fact, electro-magnets, as they are called, are weight for weight very much more powerful than permanent steel magnets (Fig. 60). Few things are more remarkable in this connection than the rapidity with which soft iron may be thus magnetised and de-magnetised. The ordinary trembling electric bell, so largely used now, supplies audible evidence of the fact; every time the vibrating hammer strikes the bell, an electro-magnet is excited, and loses its magnetism. In some dynamo-machines also these changes take place many hundred times in a minute.

The practical applications of this fact are

almost innumerable, and are daily increasing in
number.   The elaborate systems of fire-alarms,
which have been in operation in American cities
for the last quarter of a century, but are only now

Fig. 60.

beginning to be adopted in England, consist
almost entirely of electro-magnetic signalling
appliances.   In Boston it is in the power of any
person to give an alarm of fire by simply pulling
down a handle in one of the very numerous street-
boxes, when, without any further human inter-

vention, automatic machinery causes the church
bells all over the city to ring out the number of
the district whence the alarm came. So perfect
also is the fire-drill, not only in Boston, but in
nearly every large town in the Union, and so much
assistance is given by electro-magnetic arrangements, that the author has seen naked horses
loosened from their stalls, harnessed, and started
off with a steam fire-engine, in less than ten
seconds from the receipt of the signal. By an
ingenious combination of a thermometer with
this system, fires are made to signal themselves
automatically, an arrangement found sufficiently
effective to justify a considerable reduction in
insurance premiums. Similar arrangements are
employed for the protection of banks, safes, and
private dwellings from burglars, for the regu-
lation of clocks, the automatic registration of
certain occurrences, and a variety of somewhat
similar purposes.*

The system of telegraphic signalling most
largely employed upon land lines, depends upon
the attraction of an iron bar, or keeper, by an
electro-magnet, magnetised by a current sent from
the distant station. In the older instrument
known as the Morse recorder (Fig. 61), of which

---

* Among the various practical applications alluded to in the
lectures in Lancashire, was a very ingenious electric stop-motion
applied to "drawing-frames" in a cotton spinning-mill, the inven-
tion of Mr. John Bullough of Accrington, the action of which
depended on the insulating properties of cotton.

there are still about 40,000 in use on the European
continent, this keeper, *c, c*, is placed at one end
of a lever, *d, d*, to the other end of which is
attached a style or ink-pen, under which a continuous strip of paper is made to pass by clockwork, *g*. The attraction downwards of the keeper,

FIG. 61.

by the electro-magnet, *b, b*, presses the pen up
against the moving paper, and a mark is made,
the length of which depends on the time for which
the current is allowed to pass.   In the Morse
alphabet, two signals, only a (.) and a (—) are
employed, by combinations of which all letters
are signalled, no single letter requiring more than
four signals, and the most frequently used let-

ters being represented by the simplest signals;
thus—

M   A   G   N   E   T

—— —— · —— —— ——· —— ·   ·   ——

The Morse recorder thus gives visible signals,
of which a record is kept. The English and
American systems, however, have for the last
fourteen years largely employed an instrument
known as the Sounder, which is simply a recorder deprived of its recording apparatus. The
interval between two attractions of the keeper,
each of which makes a sharp " click," may
be either long or short, like a minim and
crotchet in music, corresponding to the dash
and the dot of the Morse alphabet. The signals
made in this way are audible and non-recording;
but the alphabet, once learnt, is not readily
forgotten.

Submarine cables consist essentially of a
copper conducting wire, closely and carefully
surrounded with gutta-percha as an insulator,
and the whole protected by an outer coating of
hemp and steel wire. The signalling through
them is carried on either by a very delicate
galvanometer, to the needle of which is attached
a delicate mirror on which a beam of light falls,
the movements of the needle being indicated
by the oscillations of the reflected spot of light,
or by a complicated instrument known as Sir W.

K

Thomson's Siphon recorder. There are now 80,000 miles of submarine cable at work, costing £30,000,000. A fleet of twenty-nine ships is employed in laying, watching, and repairing these cables, of which there are nine across the Atlantic alone, the last one having been laid in twelve days without hitch or stoppage. Since the purchase of inland telegraph lines by the British Government (1870), the average of 126,000 messages per week has grown to 603,000, and the "Press messages" have increased from about 5,000 words per day to very nearly 1,000,000.* According to the Postmaster-General the introduction of sixpenny telegrams involves the erection of 15,000 additional miles of wire in the British Islands.

Having seen how magnetism may be excited by electricity, we will now turn our attention to the reverse change—the development of electricity from magnetism—which is the foundation of all the modes of converting mechanical into electrical energy. This discovery, which is of superlative importance, was made by Faraday in October, 1831, quietly working experimentally in his laboratory in the Royal Institution in London, solely in the pursuit of truth, without any thought of the possible

---

* *Vide* lecture on the "Progress of Telegraphy," at the Institution of Civil Engineers, by Mr. W. H. Preece, F.R.S., reported in *Nature*, Feb. 22, 1883.

consequences that might ultimately arise from
his results. It is thus described in his own
words: *—

"A cylindrical bar-magnet, three-quarters of
an inch in diameter, and eight and a half inches
in length, had one end just inserted into the end
of the helix cylinder (220
feet long); then it was
quickly thrust in the whole

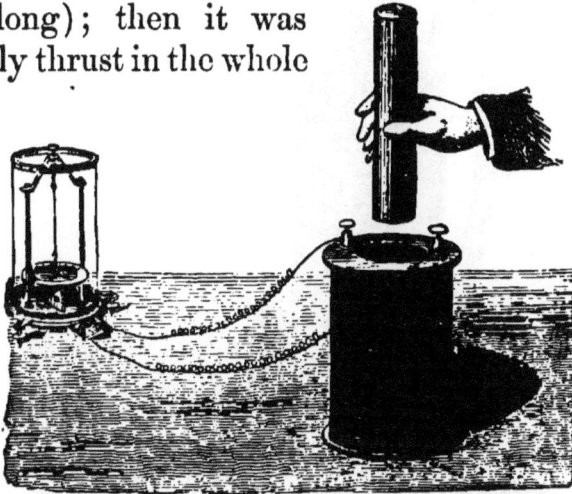

FIG. 62.

length, and the *galvanometer* needle moved, then
pulled out, and again the *needle moved*, but in the
opposite direction. This effect was repeated
every time the magnet was put in or out, and
therefore a wave of electricity was so produced
from *mere approximation of a magnet*, and not from
its formation *in situ*."

* Dr. Bence Jones's "Life and Letters of Faraday," Vol. II.
p. 5.
K 2

Fig. 62 shows this fundamental experiment, which is of superlative importance. The ends of a hollow coil of insulated wire are con-

FIG. 63.

nected with a galvanometer, and the insertion of the magnet, into the coil, or its withdrawal therefrom, *i.e.*, the motion of a magnetic field in the neighbourhood of a conducting wire (or *vice versâ*), produces a momentary current of electricity. Evidently, then, the rapid oscillation of this bar-magnet would give rise to a rapid succession of electric currents alternately in opposite directions. Similar effects, but of greater intensity, are produced by the rapid rotation of coils of wire in front of the poles of a permanent horse-shoe magnet. Such an arrangement is often known as a magneto-electric machine (Fig. 63), in which two coils of wire, t and t', revolve on a horizontal axis, f, in front of the poles of a magnet, A, B. Small machines of this kind are commonly sold by opticians, and are used for giving shocks for medical purposes. The first machine built for thus converting mechanical into electrical energy on a large scale, was simply a repetition of these parts, and one of them, driven by a steam-engine, was used for some years to produce the electric light in a lighthouse on the coast of Kent. The term "armature" is given to the rotating coils of wire, and in 1857 Sir W. Siemens devised one of peculiar form, in which the wire is wound lengthways on a spindle, for the production of the "field" in which the armature revolved. The next step was made by Mr. Wilde, who employed electro-magnets (excited by the current produced by a small auxiliary magneto-electric machine), which, as we have seen, are, weight for weight, very much more powerful than permanent steel-magnets.

The dynamo-machine in its present form, however, owes its origin to a suggestion made independently, and at the same time (1867), by Sir W. Siemens and Sir Charles Wheatstone. In it no extraneous source of electricity is needed to

Fig. 64.

keep up the current in the electro-magnet, which is excited either by the whole or by a portion of the current produced when the armature begins to rotate, since matters are so arranged that the coils of wire in the armature and those round the large electro or "field" magnets, all form part of one continuous circuit (Fig. 64). It will be

remembered that all iron is more or less magnetic, owing, in great part, to the influence
of the earth, and when the armature begins to
revolve, this residual magnetism in the cores of
the electro-magnets induces feeble currents in
the armature ; these currents then circulate
round the field-magnets, making them much
stronger, and they in their turn re-act upon the
armature, inducing stronger currents in its
coils. After a few moments' revolution, therefore, during which the machine becomes increasingly difficult to drive, its full power is
exerted, and the mechanical energy put in at
one end can be drawn off as electrical energy
at the other. The speed of rotation most suitable for the armature depends upon the particular construction of each machine, and
ordinarily varies from 600 per minute to four
times as much.

Such being the principles of the dynamo-
machine, it is almost needless to add that they
are capable of being carried out in practice in a
great variety of ways of greater or less efficiency,
according to the particular kind of work which
the machine is intended to do. In the present
transitional state of our knowledge, it may
safely be said that there is no such thing yet
as a " best dynamo." It may, however, be
well to point out in this connection what a
very efficient machine it is in comparison with

the steam-engine. It was shown in Chap. II.
that a steam-engine only yielded about one-
fourth of the total quantity of heat in it as
available for work. In the conversion of me-

FIG. 65.

chanical into electrical energy by the dynamo,
not three-fourths but only one-seventh of the
energy put into it is lost, or, in other words,
a good dynamo-machine will give out as electrical energy 85 per cent. of the energy put into
it in the mechanical form.

The various types of dynamo-machines are

usually known by their inventor's names. The
first practical one, still very largely used, is
called the "Gramme" (Fig. 64), in which the
wire of the armature is wound upon a ring, the
principle of which was invented some years previously by Paccinotti. In the "Siemens'"
machine (Fig. 65) the field-magnets are flat,
and connected by pole pieces, and the wire is
wound lengthways on the armature. In the
"Brush" the wire is wound in sections on a
ring, with intervening spaces, intended partly to
check the tendency to waste of energy as heat.
In the "Edison" the field-magnets are of
enormous length, and the armature resembles
that of the "Gramme;" and in the "Ferranti"
the armature is of copper ribbon, rotated in
a very powerful field. The largest machine
yet built is the "Gordon," weighing twenty-
two tons, and capable of supplying energy
enough for 7,000 or 8,000 incandescence lamps.
It was designed in order to do the work of a
small gas works, and to show how much more
cheaply electricity could be produced when the
process was carried out on a comparatively large
scale.*

A very important division of dynamo-
machines is into those that supply alternating
currents—*i.e.*, currents alternately in oppo-

* *Vide* Mr. Gordon's lecture, printed in the "Journal of the
Society of Arts," upon "Electric Lighting,"

sitë directions, as by the oscillation of the bar-magnet in Faraday's original experiment, and those that supply continuous currents in one direction, this being effected by suitable com-mutators for reversing the direction of half the currents. For many systems of electric lighting the former are preferable; but for the economical electric transmission of power, a continuous current machine is almost indis-pensable.

It was stated in the last chapter that the production of light from electric energy was due to some resistance interposed in the pas-sage of the current. In the "arc system" this resistance is given by two carbon-points and a thin stratum of air between them, the thickness of which stratum varies with the intensity of the current, but must be kept constant for the same current if the light is to be steady. The discharge between the points (Fig. 66) is accompanied by an actual transport of particles of carbon from the posi-tive to the negative pole, and as the carbon points are intensely heated they oxidise, and burn away with the production of carbonic acid (Chap. III.).

The proportion of light produced by this combustion is insignificant (and not essential, since the luminous arc can be maintained in a vacuum or under water); but the fact of

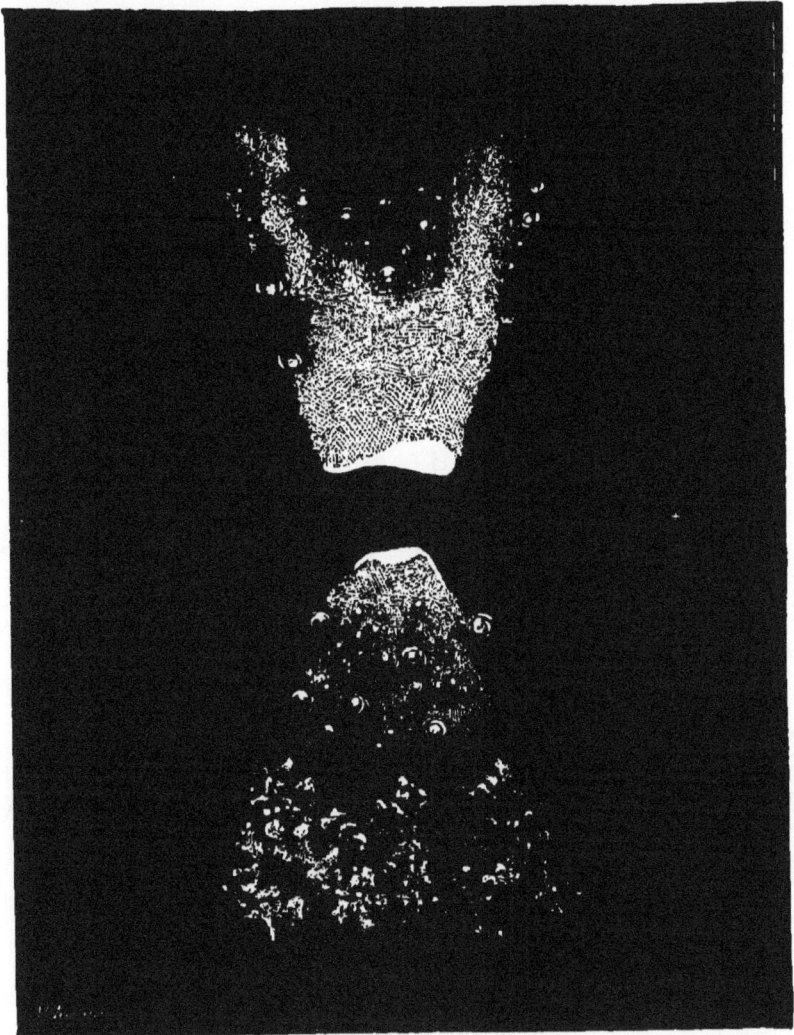

Fig. 66.

the combustion necessitates some contrivance
for keeping the points at the proper distance
apart, since they gradually waste away. This
is the object sought to be attained by the
innumerable arc-lamps, or regulators, before
the public, the general principle of the majority of which is, the mutually-opposing action
of a spring (or clockwork) which tends to
bring the points together, and of an electro-
magnet operated by the current that has passed
through the points, which tends to draw them
asunder. When their distance increases the
resistance is higher, less electricity circulates in
the magnet, and as its withdrawing influence is
lessened, the spring causes the points to approach,
and *vice versâ.*

In other lamps gravity tends to bring the
points together, while a similar application of
electro-magnetism keeps them apart. In the
so-called electric candles, such as Jablochkoff's,
the two pencils of carbon are placed parallel to
each other, and separated by some china-clay
or other infusible insulating material. Since the
pencils are always at the same distance apart they
require no machinery to regulate their distance,
but they can only be used with alternating current
dynamos.

In the incandescence, or glow, system of
lighting, with which the names of Swan and
Edison are honourably associated as the practical

pioneers, who were succeeded later by Maxim and Lane-Fox, the source of light is a continuous filament or narrow ribbon of carbon, F, placed in a glass globe, A, from which all air has been exhausted (Fig. 67). The ends of the carbon filament are attached, G, G, to platinum wires, E E, which are fused into the substance of the glass in their passage through it, and serve to convey the current to the carbon. Should the globe be accidentally broken, the lamp is instantly extinguished without setting alight to anything, because the access of air to the white hot carbon consumes it almost instantaneously. Swan's filaments are prepared by heating to red-

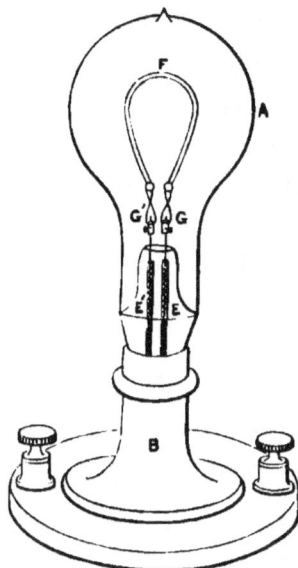

FIG. 67.

ness, out of contact of air, slips of parchment paper; Edison's by similarly treating bamboo fibre. There is no difference in principle, but only in detail, between the lamps of the various makers named above; the heat and light are in all cases due to the resistance offered to the passage of the current by the carbon filament. In the manufacture of some of these lamps, great pains are taken to adjust the resistance of each filament to an exact quantity.

Although arc-lighting is well adapted for large public rooms, railway stations, factories, and large open spaces, it is not suitable for dwelling-houses, owing to its great intensity, which cannot be diminished below a certain point, and at the same time kept steady. For such purposes, the incandescence system is eminently suited. The best example of this at present (1883) is seen in the Savoy Theatre in London, where 1,200 lamps are arranged in six circuits of 200 each, 824 of them being used to light the stage.* Several of the highest class of ocean passenger steam-ships are now lighted in this way, the freedom from risk of fire, and the absence of any noxious products of combustion, such as carbonic acid, &c., rendering the system peculiarly advantageous for this purpose, as it is also for libraries, picture-galleries, silversmiths'-shops, &c., the contents of which are seriously injured by the sulphuric

* The following statistics of the Electric Lighting at the Fisheries Exhibition (first shown on June 26th) may be interesting. As many as 4,000 incandescence lamps and nearly 300 arc lamps are employed, the latter varying from 280 to 500 candle-power. The amount of real, not nominal, candle-power thus distributed through twenty-five miles of wire is estimated at more than half-a-million of candles, and to produce it, 1,100 indicated horse-power is used up in the steam-engines, of which there are more than fifty, each driving its own dynamo-machine.

and other acids arising from the combustion of gas.

It may be interesting to note here the difference in the amount of mechanical energy necessary to produce the same amount of light by these two systems of electric lighting. The expenditure of one horse-power upon a good dynamo and arc-lamp will give in the latter light equal to from 1,000 to 1,200 candles. The same energy expended upon incandescence lamps, will only yield about 180-candle-power light, *i.e.*, will only keep going from 8 to 9 of Swan's standard 20-candle-power lamps.

Into the comparative cost of lighting by gas and by electricity, it is beyond the scope of this volume to enter, since it involves questions of the cost of obtaining power, which is so very variable, especially when the supplies of natural energy running to waste in waterfalls, wind, and tide, are taken into consideration ; supplies which, through the agency of the dynamo-machine and the storage batteries (Chap. III.), may now be rendered available for the production of electrical energy.

It may, however, be worth notice, as an interesting illustration of various transformations of energy, that if a given volume of gas be burnt in the ordinary way, and the amount and duration of the light be noticed, and if an equal volume of the same gas be burnt in an Otto gas-engine

(p. 81), which drives a dynamo-machine supplying an arc-lamp, the amount of light in the
second case will be about seven times as much
as in the first, owing in great measure to the
fact that so little energy is wasted in producing the heat which always accompanies gas-
lighting.*

It may be stated broadly that when equal inten-
sities of light (*i.e.*, the same number of " candle-
power") are produced, arc-lighting is, under almost
any practical set of conditions, cheaper than gas-
lighting. The same, however, cannot be said
of the only form of electric light available for
domestic use, viz., the incandescence system,
since it requires at least five times as much
mechanical energy as an arc-lamp for the
same amount of light. The comparison of
cost of electric and gas-lighting varies with
the local conditions in each town, such as the.
price paid there for gas, the price of steam
coal, and the presence or otherwise of waste
water-power ; hence each case can only be
dealt with on its own merits; in making the
comparison, however, the advantages of electric
lighting alluded to on p. 158 should not be overlooked.

Having now a clear conception of the way in
which mechanical energy is changed into electrical,

* *Vide* the Modern Applications of Electricity, by E. Hospitalier,
translated by Julius Maier, pp. 264, 265.   (Kegan Paul & Co.)

let us consider the reverse change, viz., how
to produce mechanical energy from electrical,
or, in colloquial language, the use of electricity as a motive power. As long as electricity could only be produced by chemical, and
therefore costly, means (the relative costs of
the chemical and mechanical methods being as
120 to 1; *vide* p. 110), this subject attracted
comparatively very little attention from the
so - called practical men, although scientific
workers had paid much attention to it, and
had devised various pieces of apparatus for
effecting it.

As already pointed out, Oerstedt's discovery
(p. 137) of the deflection of a magnetic needle
when an electric current passed near it, was
the first instance of the electrical production
of mechanical motion. The discovery of the
electro-magnet, and of the great attractive force
it exerted on its keeper, gave a great im-
petus to the construction of electro-magnetic
engines of various kinds, in which mechanical
motion was usually produced by the attraction
for a piece of iron of magnets alternately excited. In 1839 Jacobi propelled a boat along
the river Neva with such an engine of about
one horse-power, worked by a battery of sixty-
four large Daniell cells (p. 104). The electric
launch which was recently propelled on the
Thames had this very great advantage over it

L

in point of cost, viz., that the electricity had
been produced mechanically by a dynamo-
machine, and the energy had been stored electrically in some secondary batteries (pp. 122–126)
until required for use.

In the present day, however, the transformation of energy necessary for the employment of
electricity as a motive power, is effected by the
agency of the dynamo-machine. One of the
keenest intellects of the age, the late Prof. Clerk
Maxwell, of Cambridge, replying to an enquiry
as to what he considered the greatest discovery
of the last twenty-five years, said, " That the
Gramme machine is reversible." In other words,
a dynamo-machine (for the remark applies to
them all, though Gramme's was the best known
at the time) may be worked in both directions,
forwards and backwards. If supplied with
mechanical energy, it will give out electrical,
but *if supplied with electrical, it will give out
mechanical.*

This is the principle upon which all the uses of
electricity as a motive power, and the electrical
transmission of power to a distance, depend. To
make an electric current set a machine in motion,
then, it is necessary to interpose between the ends
of the conducting wire and the driving pulley of
the machine, a small dynamo-machine worked
backwards. To such small machines the term
electric motor is usually given, but it must be

distinctly understood that the same machines are
capable of being used both as electric generators
and as motors. A large central generator may
(and at the present time in New York *does*) supply
electrical energy to a large number of small
motors (Fig. 68), all connected with it by wires,
each of which may work a separate machine. This
is practically carried out by Sir. W. Siemens on

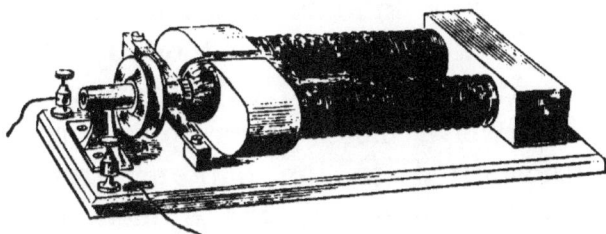

Fig. 68.

his own farm, a saw-mill, a set of pumps, a chaff-
cutter and other machines, in different parts
of the farm, all being driven and controlled from
the dynamo and central steam-engine. When the
comparative cost of working large and small steam-
engines is taken into account, this mode of "dis-
tributing power" electrically will be found to be
very economical. It was in this sense that the late
Dr. Spottiswoode, President of the Royal Society,
recently expressed his belief that "before long
electricity would prove the poor man's friend,"
enabling small amounts of motive power to be
brought to his own door for industrial purposes,

L 2

instead of his being obliged to go to the factory
for it.   By these means electric railroads and
tram-roads are worked, the energy being in some
cases drawn from storage batteries (previously
charged by a dynamo-machine) carried on the
vehicle, and in others transmitted along a wire
or cable, between which and the car there is
maintained a "rolling contact."*   In the Paris
electrical exhibition of 1881 there were examples
of almost every kind of mechanical operation
(even to rock-boring) carried on by electrical
energy.   At the electrical exhibition at Munich
in 1882, twelve horse-power of energy was
daily transmitted to the building from some
waterfalls five miles off, the dynamo-machines
there being turned by large turbines, which
are peculiarly fitted for such work, from their
rapid speed.   In all cases, however, where
machines are apparently "driven by electricity,"
it is important to remember that there is (or
has been) mechanical energy at their back, in
some shape.

This electrical transmission of power, again,

---

* The special advantage, speaking in general terms, of electric
railroads, is the very small weight of the motor, as compared with
that of a steam locomotive.   At present, all bridges, embankments,
&c., have to be made two or three times as strong as they need be,
if the weight of the train only, without the locomotive, had to be
considered.   In particular cases, such as badly ventilated underground railways, long tunnels, &c., electric motors possess other
obvious advantages, since no noxious products of combustion are
given off by them.

affords the means, either with or without the intervention of secondary batteries, of utilising
sources of natural energy that hitherto have
been running to waste, and it is only in this
limited sense that electricity may be considered
as a new force, or "the coming force," as it is so
often erroneously called, and only thus can it be
regarded as an *addition* to the means already
possessed by man of doing work.

One instance of the amount of this "wasted
energy" may be cited. At the site of the
Severn Tunnel, ten miles from Bristol, the
river is two and a half miles wide, and the
average rise of tide is fifty feet. If the
average rate of flow across this section were
only one mile per hour (which is certainly a
very low estimate), 100,000 horse-power could
be utilised; the market value of this at present
is something like £1,000,000 per year, which
is now allowed to be wasted. The necessary engineering works would be very simple,
though on a large scale, and the action could
be made continuous, and independent of the
state of the tide, without using electric accumulators.

Probably the best example of this use of waste-
energy is the case of waterfalls before alluded to,
whose power can now be transmitted electrically
to distant cities, and there used either for lighting or for mechanical work. In his address to

" Section A " of the British Association for the Advancement of Science at its meeting in 1881, Sir W. Thomson showed that a copper wire of half an inch diameter charged at Niagara with energy equal to 26,250 horse - power, would give out at a distance of 300 miles (a greater distance than New York, Boston, Montreal, or Philadelphia) energy equal to 21,000 horse-power, and that the cost of this length of wire at 8d. per pound is £37,000, the interest on which at 5 per cent is £1,850 per year.

The numerical relation between heat-energy and work was pointed out in Chap. II., 772 foot-pounds being the number that expresses the mechanical equivalent of heat. The same kind of exact relation exists between electrical energy and work. The mode of measuring electrical energy by galvanometers and voltameters of various kinds has been indicated generally in the preceding pages. In measuring the energy that can be got out of a given volume of water used in hydraulic machinery it is necessary to know not only the quantity of water used, but also the pressure or "head" at which it is supplied. The same thing is true of electricity ; both the pressure, *i.e.*, the electro-motive force (p. 106), and the quantity, must be known.

Now, the first of these (the E.M.F.) is measured

in *Volts*, and the quantity in *Ampères*, and hence
a given quantity of electrical energy is measured
in *volt-ampères*. One horse-power is equal to 746
volt-ampères, or, to use the single word proposed
by Sir W. Siemens, to 746 *watts*. So practical
a question have these measures become, that in
many of the Provisional Orders for electric lighting purposes now before Parliament, the proposal
is made* to call by the name "one unit" the
energy contained in a current of 1,000 ampères
flowing under an electro-motive force of one volt
for one hour, and for this "unit" of electrical
energy it is proposed to charge sevenpence.
Now, as "one unit" of energy is 1,000 volt-
ampères (or 1,000 watts), and as 746 watts are
equal to one horse-power, it follows that the *price
of energy delivered electrically* is about 5⅓d. per
horse-power per hour.†

Another illustration on a very much smaller
scale, though perhaps still more interesting, of
the transformation of mechanical into electrical
energy and *vice versâ*, is afforded by that wonderful little instrument for transmitting articulate
speech to long distances, the Bell Telephone
(Fig. 69). Like the dynamo-machine, it is a

---

* *Vide* "Nature," No. 695, Vol. XXVII., p. 385.

† In connection with the general subject of the Transmission of
Energy, it may be interesting to note that Professor Osborne Reynolds, F.R.S., recently expressed his opinion before the Society of
Arts that it would be a profitable speculation to supply stored
energy in the shape of air compressed in steel bottles, at the rate
of 3d. per million foot-pounds.

reversible engine, and may be worked in either
direction, *i.e.*, it may be used both to receive
and to transmit messages. It consists of a steel
bar-magnet (*d*), round one pole of which is a
coil of insulated wire (*a*), the ends of which
(*e, e*) are connected (*f, f*) with the line running
to the other station and with the earth-circuit.
Immediately in front of the same pole,
but not actually touching it, is a circular

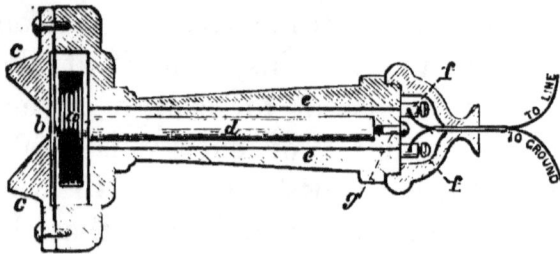

FIG. 69.

diaphragm, or disc, of thin sheet-iron (*b*),
supported only at its edge, close behind a
mouthpiece (*c, c*).

As is well known, sound travels through the
air in waves; the number of these waves that
strike the ear in a second determines the pitch
of the sound (*i.e.*, the musical note), and their
size or height influences the *timbre* or quality
of the sound. When these waves in the air
strike upon the drum of the ear they cause the
sensation of sound; when they strike on the diaphragm of the telephone they cause it to vibrate.

The vibrations of this iron disc close to the pole of the magnet alter its " field " continually, and these alterations excite very feeble currents of electricity in the coil surrounding that pole; in this manner the atmospheric waves are converted into electrical waves, or undulatory currents, as Graham Bell calls them. On their arrival at the distant station these currents enter a precisely similar instrument; circulating round its coil they produce a succession of feeble changes in its magnetic field, each one corresponding to a similar change in the transmitting instrument. These feeble changes in its field cause the magnet to attract and repel its disc, i.e., to set it vibrating ; these mechanical vibrations are at once communicated to the air, and, falling on the ear of the listener, produce the sensation of sound. With a pair of such instruments only it is possible not merely to converse, but to recognise a friend's voice, at a distance of several miles.

A telephone, then, converts atmospheric into electrical waves, and vice versâ. Its currents are too feeble to overcome the resistance of a great length of wire, but if in any way an undulatory movement could be set up in a stronger current, capable of travelling a great distance, the telephone would convert these undulations into sound, and thus the distance-limits to conversation could be enormously increased. This is

actually effected by the microphone of Professor
D. E. Hughes (Fig. 70), in which two or three
pieces of carbon lying together in "loose contact" are put into the circuit of a few cells of
a battery, in which circuit also a telephone is
included.   As long as these pieces of carbon are
perfectly still, a constant current passes, and no
sound is heard in the telephone.   The smallest
vibration (even that produced by a fly walking)

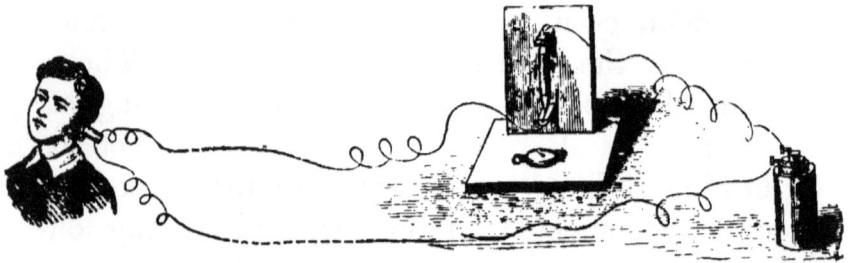

Fig. 70.

in the neighbourhood of the carbons, however,
whether produced by speech or in any other way,
shakes them and alters the contact, causing a
variation, or undulation, in the current, which
the telephone renders audible.   All telephonic
transmitters are simply combinations of micro-
phones, and with their aid conversation is possible up to distances exceeding 200 miles, and
the sound of music can be transmitted above
500 miles.   Conversation has taken place also
recently between New York and Chicago, a
distance exceeding 1,000 miles; in this case,

however, an ordinary telegraph wire was not employed, but a steel wire coated electrically with copper.

A more remarkable feat still, perhaps, also achieved by Graham Bell, and also illustrating the transformations of energy, is the transmission of articulate speech electrically, without the intervention of any conducting wire, by an instrument called the Photophone. It depends for its action upon the very curious fact that the electric resistance, or power of conducting electricity, of the rare element selenium, is diminished by the radiant energy of light. If therefore a surface of selenium be included in an electric circuit with a battery and a telephone, no sound will be audible in the telephone as long as there is no change in the intensity of the light falling upon the selenium, but any such change alters the resistance of the selenium, and causes the current to vary in strength, or to become undulatory in character, producing sounds in the telephone. The necessary changes in the light are produced at the transmitting station, by speaking to the back of a thin mirror, so adjusted as to reflect a beam of light upon a parabolic (curved) mirror, at the receiving station, which concentrates the beams on the selenium cell; the words spoken at the transmitting station cause the mirror to vibrate, hence the beam falling on the selenium vibrates also, undulations are thus set up in

the electric current, and these are translated by
the telephone into articulate speech ! There are
probably few, if any, more curious and interesting examples of the transformation of energy
than this.

CHAPTER VI.

ENERGY IN ORGANIC NATURE.

At the end of Chap. III. a brief reference was made to the power of that form of radiant energy which we call Light, to do chemical work in the shape of effecting the separation of certain chemical substances from each other ; and the decomposition in that way of the salts of silver, upon which the art of photography is based, was adduced as an instance. It was also stated in passing, that under the influence of the radiant energy of the sun, growing plants decomposed the carbonic acid existing in the air (pp. 74 and 86) setting free oxygen, and storing up carbon in their own substance. This process we are now to consider a little more in detail, and for this purpose it will be well to glance first, very briefly, at the life-history of some familiar plant, that of the oak from the acorn, for example.

If an acorn be slit lengthways down the middle, the greater part of its contents will be seen to be a yellowish white substance, but at one end there is a little thing about the size of a pin's head, which contains the germ of the future plant, a rootlet, and a rudimentary stem. The rest of the acorn, the yellowish white substance, is nothing more nor less than a store of nourishment ready prepared for the plantlet, until it is able, as it were, to work for its own living. When the acorn falls into the ground, so long as the earth is cold and dry, it lies as though dead, or like a person in a trance; but as soon as the sun waves pierce down into the earth, their energy wakes up the plantlet, agitates the particles of matter in it, and causes them to combine with the particles of ready-made food—starch, oils, sugar, and albuminoids—which are laid up for the plantlet in the thick seed-leaves in which it is buried (Fig. 71).

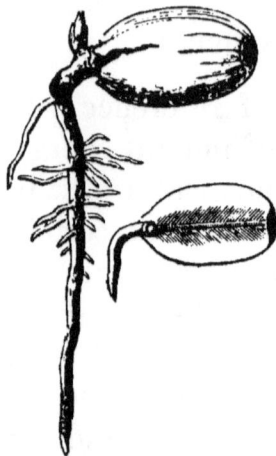

Fig. 71.

In this way, then, does the baby-plant spend the first stage of its existence, and the root and stem are developed at the expense of the previously prepared food carefully elaborated for it by the full-grown oak-tree. By the time, however, that this store is exhausted, portions of the plant have appeared above the ground and are exposed to light, and as soon as this occurs, those portions have become capable of preparing for themselves their own food, while the radiant solar energy works through them, as it were.

Careful experiments have clearly shown that the green .surfaces of growing plants absorb carbonic acid from the air, and give out oxygen, at the same time that they increase in weight from the deposition of carbon in their tissues, which are chiefly formed by the combination of this carbon with hydrogen and oxygen derived from the water absorbed by the roots. Woody fibre consists chemically of thirty-six parts by weight of carbon, five of hydrogen, and forty of oxygen.

The evolution of oxygen by growing plants may be readily shown by putting some freshly gathered green leaves (laurel leaves answer well) into a tumbler full of, and inverted over, water; when the sun shines on this, bubbles of oxygen are given off by the leaves in abundance. In an aquarium also, a stream of such bubbles may often be noticed rising from a broken stem of the water-plants Anacharis or Valisneria.

At the same time that this process goes on, the root fibres of the plant absorb water, and, dissolved therein, various mineral substances, as potash, &c., and especially substances containing nitrogen, all necessary for its proper nourishment. Much of the nitrogen reaches the plant in the shape of ammonia (composed of fourteen parts by weight of nitrogen and three of hydrogen), from which the plant prepares those elaborate nitrogenous compounds, which, as we shall presently see, form an essential part of the food of man.

It is very interesting to note that the radiant energy of the electric light is nearly, if not quite, as effective in promoting these changes in the plant, and this elaboration of its food, as that of sunlight. This subject has been carefully investigated by Sir W. Siemens, who has been able, by a judicious combination of hot-house warmth and electric light, to control at will the time of the flowering of plants and the ripening of fruits, and such flowers and fruit have been pronounced by competent judges to be equal in all respects to those grown in the ordinary way.

In November, 1881, two plots of precisely similar ground on Sir W. Siemens's farm were sown with oats, wheat, and barley, the same seed being used in each case. One plot was not interfered with in any way; the other was lighted by the electric light from sunset to sunrise, until May, 1882, when ears were cut from each plot. Fig. 72, from a photograph kindly permitted by Sir W. Siemens, shows the result in each case. The three long ears were grown in the portion electrically lighted at night, the three short ears in the other; the difference is greatest in the case of the oats, and least in that of the wheat. A very remarkable feature in these experiments, first pointed out by Sir W. Thomson, and fully confirmed by Sir W. Siemens, is, that no artificial light will effect this decompo-

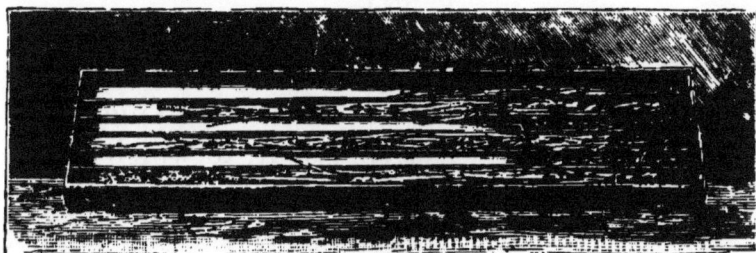

FIG. 72.

sition of carbonic acid in growing plants, unless the temperature of the source from which it comes is above that necessary for the dissociation of carbonic acid, viz., 4,000° Fahr. (p. 85). The electric light is, as yet, the only artificial light that fulfils these conditions. Another inference from the experiment is, that the period of rest, hitherto considered necessary for plants during their growth, is not nearly so essential to them as was generally supposed.

The question now arises, what part of the plant is it that effects this decomposition? The

M

flesh of all plants and animals is made up of
little (transparent) bags of fluid growing one
against another, called cells (Fig. 73). In pith
they are round, large, and easily seen; in stalks
of plants they are usually long, and overlap each
other; in other parts of the plant they vary much,
both in shape and
size. The fluid which
these cells contain is
a sticky liquid with
minute grains con-
stantly moving about
in it, to which liquid
the name "proto-
plasm," or *first form* of
life, is given. In the
growth of the plant
these cells multiply
by being sub-divid-
ed, the protoplasm of
one cell dividing into two parts and building
up a wall between them, one cell thus becoming
two. In the simpler plants and animals, composed of single cells or of one kind of cell, one
cell has to do a great many kinds of work, per-
forming all the different functions of the organism; but the more highly organised a plant (or
animal) is, the more are these different duties or
functions divided among various kinds of cells,

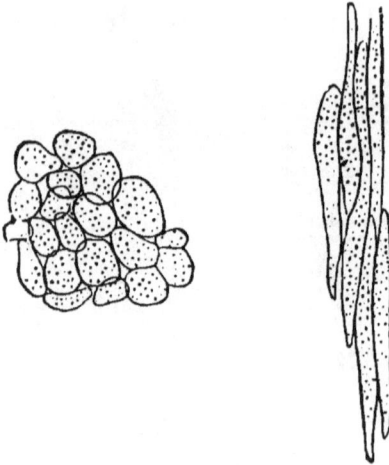

Fig. 73.

only one duty being allotted to each. Thus (Fig.
74) one set of cells will secrete starch (*st*), another
oil (*a*), another nitrogenous
compounds, a fourth will
decompose carbonic acid, a
fifth has to do with respiration, and so on. *Why* one
cell is able to do many
kinds of work, while ano-
ther, apparently precisely
similar, is only able to do
its own special kind, is a
question about which we are as ignorant (or

Fig. 74.

perhaps more so) as we are of
the nature of the various forms
of energy around us, which are
unknown apart from matter.

The protoplasm in the cells
that decompose carbonic acid
(Fig. 75) is of a beautiful green
tint; to it the name *chlorophyll* is
given. It absorbs all the solar
rays except the green ones, which
it throws back again, producing
in our eyes the sensation of a
green colour.

Under the influence of solar
energy, then, carbon and oxygen
are separated from each other.

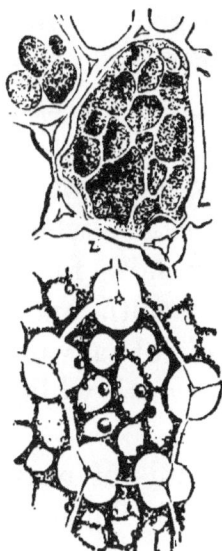

Fig. 75.

In Chap. III. it was shown that when chemical

M 2

attraction exerted itself between these two elements, heat and light were the result. Hence,
when wood is burned (and, as we shall presently see, when vegetable food is " burned,
i.e., oxidised, in the animal body), the energy
thus generated is simply a *reproduction of the
solar energy* which separated the carbon and
oxygen during the growth of the plant, just
as in the secondary battery (pp. 122–126) the
electrical energy drawn off when the cell is
discharged, is due to the re-union of the
hydrogen and oxygen which had been previously separated from each other by the
decomposing action of the current employed
to charge the cell. As was remarked in the
detailed explanation of that, there is a complete
analogy between these two cases of the storage
of energy.

Now, this line of argument may be carried
back into those distant geological ages when
the mighty forests flourished which now form
coal, that great store of our national wealth, and,
regarding our coalfields in the light of a vast
store of potential energy, it may be worth while
briefly to trace their history.

A careful examination of many kinds of coal
shows that it will split into layers, and if the coal
be examined in the mine it will be found to lie
between beds of a hard mineral called shale,
which splits into layers much more readily than

the coal itself does. In this shale are frequently found the impressions of various portions of plants, such as those seen in Fig. 76. Moreover,

Fig. 76.

in the underclay, or floor, of the coal-bed, are found curious fossils (Fig. 77), which investiga-

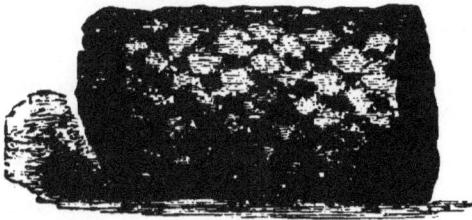

Fig. 77.

tion has shown to be the underground root-stems of some of the trees, the remains of whose trunks were found in the shale and coal. Speaking in technical language, the *Stigmaria* has been shown to be the root-stem of the *Sigillaria*. There are found also occasionally imbedded in the coal, round nodules of stone like cannon-balls, which, when cut across (Fig. 78) in transparent

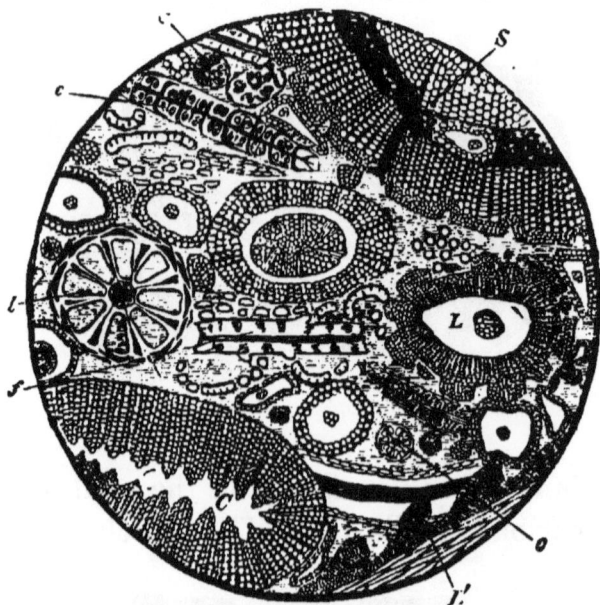

FIG. 78.—CONTENTS OF A COAL-BALL.   (*Carruthers.*)*

S, Stem of Sigillaria cut across.   *L*, Stem of Lepidodendron cut across. *L'*, Stem of Lepidodendron cut lengthways.   *l*, cone of Lepidodendron (Lepidostrobus) cut across.   *C*, Stem of Calamite cut across.   *c, c, c,* Fruit of Calamite lengthways and across.   *f*, Stem of a fern with fragments of fern-leaves scattered round it.   The small round dots scattered here and there are the larger spores which have fallen out of the fruit-cones.

slices, show the leaves, stems, &c., of the coal-plants.

In the coal-forests, then, an ideal drawing of a scene in which is given in Fig. 79, there were,

* I am much indebted to Miss Arabella Buckley for allowing me to copy this figure from the original diagram of a coal-ball by Mr. Carruthers, of the British Museum.

Fig. 79.

broadly, three kinds of trees or plants—a tree
with a scaly trunk, called *Lepidodendron;* a tree
with very curious markings on its stem, called
Sigillaria; and a plant with a reed-like stem,
called *Calamites;* and associated with these grew
large tree-ferns, such as are now seen in the
forests of New Zealand and Australia, as well as
many smaller ferns.   None of these coal-forest
plants, however, bore any flowers, and when in process of time flowering plants were developed, they
began to crowd out the giants of the coal-forest
in the struggle for
existence, and the lat-
ter gradually dwindled
down, so that their re-
presentatives living in
the present day are
very   small   indeed.
The little club - moss,
Lycopodium (Fig. 80),
which grows best on
the moors in the North
of England, is the de-
scendant of the gi-
gantic Lepidodendron,
and the "horse-tail,"

FIG. 80.

or *Equisetum* (Fig. 81), is all that remains of the
Calamites.

In connection with the history of the formation of coal it is interesting to note, that the same

processes which went on during the period of
the coal-formation are now going on in a marshy
district of North Carolina, in the United States;
the plants are of course different, but the general
state of things as regards the atmosphere and
the ground is the same. In this great dismal
swamp trees and plants go on growing and
shedding their leaves year by year, and as
they fall into the water they are protected
from the decomposing
action of the air, and
accumulate layer upon
layer : the streams that
feed this swamp bring
perfectly clear water,
since all mud is fil-
tered from them by
the tangled mass of
roots, stems, &c., at
the outside edges of
the swamp; thus a
great thickness of
vegetable remains is
being formed free
from any earthy mat-
ter, producing what we
now recognise as peat.

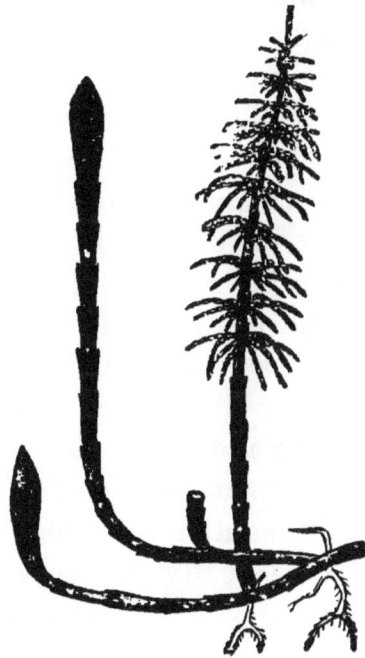

FIG. 81.

The transformation of the beds of peat into
coal was due to the various changes of level to
which they were subjected in the course of geo-

logical ages, owing partly to the "energies within
the earth" (such as volcanoes), and to the contraction of the crust of the earth as it gradually cooled
(causing cracks in its surface, which produced
mountain ranges), and partly to the wearing-
down action on the earth's surface of water, ice,
and atmospheric changes.   Owing to combinations of these various agencies, the peat-beds
gradually became covered with hundreds of feet
thickness of other material, usually rock, the
weight of which consolidated the peat, and ultimately produced coal.*   In some cases, too, the
outburst of volcanic rock near a partly-formed
coal-bed distilled portions of it, and produced
the natural mineral oils so abundant in some
parts of the world, and also gave rise to the gas
which, as is well known, often issues from cracks
in the coal, and in ill-ventilated mines produces
the disastrous results of an explosion (p. 77).

The unquestionably vegetable origin of coal
having now been shown, and the chemical changes
that occur during the growth of plants having
been pointed out, the truth of the statement that
coal is really a reservoir or store of the sun's
energy which poured down upon the earth in
past ages, will be apparent.   Great and powerful minds often see things by instinct, as it
were, or intuitively, without going through the

---

* This conversion of peat into coal can be effected in the laboratory on a small scale by very strong hydraulic pressure.

whole steps of a demonstration, and it was the intellectual perception of the mutual relation of these facts that led George Stephenson to say, in reply to a question as to what drove his locomotive, "The sun;" and on other occasions to refer to coal as "bottled sunshine."

We now come to the consideration of Energy in the animal kingdom. An exact definition of an animal as distinguished from a plant, is exceedingly difficult to frame; probably a thoroughly satisfactory one never will be found, since the tendency of all modern scientific results, not only in biology, but in other branches, is to obliterate all such strongly marked distinctions and systems of definite and self-contained classifications. Indeed there is little doubt that organisms exist which are, to all intents and purposes, plants at one period of their life-history, and animals at another. A broad distinction, however, between these two great classes of living beings, is to be found in the nature of their food, a distinction which applies at any rate to the higher animals and higher plants, and probably does to the lower ones also, although it is difficult in some cases to prove it.

Plants, as we have seen, derive their food from simple substances,—carbonic acid, ammonia, and water—from which they build up most complicated compounds. Animals, on the other hand, require that their food should be previously prepared and elaborated for them, by plants or by other animals. Plants, in fact, convert actual, radiant, moving energy, into a store of potential energy; animals, on the other hand, may be regarded as machines which convert potential into actual energy. The potential energy is supplied to the animal in its food, and this is converted, by the chemical changes within the body, into the actual energy of heat and mechanical work. These chemical changes are in the main, an oxidation of the various constituents of the food, a slow combustion of them in fact, a change of precisely the same nature as that explained in Chap. III. under the head of Combustion, although occurring more slowly, and therefore its effects are spread over a longer time, and are not so immediately evident. If a given weight of carbon be burnt in the air, or consumed in an animal's food and slowly oxidised in his body, the amount of heat (and therefore of work got out of that heat) is the same in both cases. The income of animal energy, therefore, consists in the oxidation of food into its waste products, while the expenditure takes the form of the heat necessary to the maintenance of life, and of the mechanical work done.

It has been pointed out by Helmholtz that only *one-fifth* of the energy produced by the combustion of food can be used *outside* the human body; the remaining four-fifths being required

for doing the vast amount of *internal work* necessary in the body, such as the movements of
the heart and lungs, &c., and in the maintenance
of the animal heat. The normal temperature of
the human body is 98° Fahr., and it is obvious
that the combustion of a considerable amount of
material is .necessary to keep so large a surface
and so great a bulk at a temperature several
degrees above that of the surrounding air. It is
a well known fact that when people die of
starvation, the cause of death is really *cold*—*i.e.*,
inability to maintain the heat necessary for the
vital processes; hence, when food is scarce,
much less is needed if the body be kept
warm; a fact noted by a careful Lancashire
housewife, who, during the last cotton famine,
kept the men of her household in bed and well
covered up, for about twenty hours out of the
twenty-four.

Animal heat, then, is kept up by the slow
oxidation of the carbon, hydrogen, and nitrogen
in the food that we eat, and animal work, as we
shall presently see, is derived from the same
source. "Observe," says Sir W. Armstrong in
his address on the steam-engine referred to on
page 54, "how superior the result is in Nature's
engine to what it is in ours. Nature only uses
heat of low grade, such as we find wholly un-
available. We reject our steam as useless at a
temperature which would cook the animal sub-

stance, while Nature works with a heat so mild as not to hurt the most delicate tissue. And yet, notwithstanding the greater availability of high-grade temperature, the quantity of work performed by the living engine, relative to the fuel consumed, puts the steam-engine to shame."

It will be desirable here to point out the three classes under which the various things composing our food may be grouped. The first group comprises all those articles of food in which starch and sugar are the chief components. These substances consist of carbon, hydrogen, and oxygen, but the two latter substances are present in almost exactly the proportions necessary to form water. Hence, the only energy derivable from them is due to the oxidation of the carbon which they contain. To the second group belong all oils, fat, butter, &c., substances which contain large quantities of carbon and of hydrogen, united with only a small proportion of oxygen. Hence, for equal weights, much more energy is obtained from foods of this group than from those of the previous one, and it will be remembered that the amount of heat-energy developed by the combustion of hydrogen (p. 71) is greater than that from the same weight of any other known substance. The third great group of foods contains the element nitrogen, in addition to carbon, hydrogen, and oxygen, and the purest example of it is seen in the white of an egg, a substance called albumen by the chemist; other instances are seen in the gluten of wheat (the stringy substance that remains when all the starch has been washed away from flour by a stream of water), in the fibrin of meat, the casein of cheese, and also in certain vegetables, notably lentils, peas, and beans.

When these various substances are oxidised in the animal body, by the oxygen drawn into the lungs during breathing, carbonic acid is produced from the carbon, and water from the hydrogen, and both of these products pass off chiefly in the breath. The oxidation products of the nitrogen form a rather complex substance called urea, in the getting rid of which from the body the kidneys play a very important part. The evil influence of carbonic acid when present even in small quantities in the air, and the tests for it, were pointed out in Chap. III.; but it may be interesting to note in this connection that, taking the case of an adult with an average alternation of activity and repose, about 360 cubic feet of air pass through his lungs in twenty-four hours, and as the air exhaled contains about one-twenty-fourth part of carbonic acid, it follows that about 15 cubic feet of that gas, containing about 8 oz. of solid carbon, are thrown off from the lungs of each adult in every twenty-four hours.

It was formerly thought, on the authority of

Liebig—who, although a very eminent chemist, was not a physiologist—that hard bodily work was done at the expense of the muscular tissue itself, which was actually destroyed in the process. More careful experiments, however, showed that this could not be the case, and one very curious instance may be adduced—the amount of work done inside the human body by the heart in maintaining the circulation of the blood is so great, that if it were done at the expense of the muscular tissue of the heart itself, a new heart would be required every week! Again, if Liebig's doctrine were true, whenever excessively hard work is done, there ought to be a corresponding increase in the production of urea, the oxidation-product of nitrogen, and this also is found not to be the case. It is, however, the fact that during hard work a larger supply than usual of albumenoid food is necessary, but this is probably due rather to the *rate* than to the *extent* of the chemical change which violent exercise causes in the human body.

Liebig's doctrine was completely disproved by the careful experiments of two German chemists, Fick and Wislicenus, who in their own persons were the subjects of the experiments. The work that they did was the ascent of the Faulhorn, a Swiss mountain; as its height was known, and also their own weight,

the number of foot-pounds of work done by each
in making the ascent could be readily calculated.
For a few days previously they fed themselves
upon starch, sugar, &c., avoiding all nitrogenous
compounds, and they carefully measured and
weighed the products of the oxidation of this
food ; they then found not only that there was
no notable increase in the production of urea
(the result of the oxidation of the nitrogenous
constituents of the body), but that the amount
thus oxidised only corresponded to one-fifth
part of the total work done. It was clearly
shown also that the work *was* done at the
expense of the carbon and hydrogen contained
in the food.

Dr. Frankland has investigated the same
question from a different point of view, and
has shown the amount of heat that is produced by the complete combustion or oxidation of different kinds of food, and this being
known, the mechanical equivalent of heat investigated by Joule (pp. 35, 36) enables the same
thing to be expressed in terms of mechanical
work.

Here are some of Dr. Frankland's results :—

Heat derived from the complete combustion of 1 lb. of
each of the following substances, translated into foot-tons of
work (1 lb. of water raised 1° Fah. = 772 foot-pounds = $\frac{1}{3}$rd
foot-ton nearly).

N

Wheat-flour	.	.	.	.	2,283	foot-tons.
Oatmeal	.	.	.	.	2,439	,,
Peas	.			.	2,341	,,
Potatoes		.			618	.,
Lean beef			.		885	.,
Beef-fat .				.	5,626	,.
Butter .				.	4,507	..
Mackerel		.		.	1,000	,,
Rice (Chinese)		.		.	2,330	,,

Dr. Frankland put the same thing in another way. He contrived a means for ascertaining the amount and the cost of food necessary to be completely oxidised in order to do the same amount of work as that done by Fick and Wislicenus, viz. raising a man's weight (140 lbs.) 10,000 feet high, with the following results :—

	lbs.	Price per lb.		Cost.	
		s.	d.	s.	d.
Bread .	2·345	0	1½	0	3½
Oatmeal.	1·281	0	2¼	0	3½
Potatoes .	5·068	0	1	0	5¼
Beef-fat .	0·555	0	10	0	5¼
Cheese . .	1·156	0	10	0	11½
Butter . .	0·693	1	6	1	0½
Lean beef .	3·532	1	0	3	6½
Pale ale . .	9 bottles	0	6	4	6

This table shows clearly the *real* value of various kinds of food compared with each other, provided that when taken they are completely oxidised, and properly assimilated and digested ;

but in arriving at a sound conclusion as to the practical value of any kind of food in any particular case, the two last points raised are of great importance. Into the general question of dietaries space forbids us to enter, but it may be stated generally that the daily need of the average adult requires as a minimum the assimilation of 4,900 grains of carbon and 300 grains of nitrogen, and that such quantities are found in the following dietary :—

Bread	18	ozs.
Butter	1	,,
Milk	4	,,
Bacon	2	,,
Potatoes	8	,,
Cabbage	6	,,
Cheese	$3\frac{1}{2}$	,,
Sugar	1	,,
Salt	$\frac{3}{4}$	,,
Water, alone, and in tea, coffee, beer, &c. . .	$66\frac{1}{4}$	,,
Total	6 lbs. $14\frac{1}{4}$ ozs.	

The food that forms and repairs muscle, and gives permanent capability of muscular power, must not be confounded with that which supplies the requisites for temporary activity. An examination and analysis of various public dietaries, containing the food found necessary for men doing work of various degrees of

N 2

hardness, shows clearly that while in *very* hard work the amount of hydro-carbons (such as fat, starch, sugar, &c.) has to be *doubled* as against the amounts necessary for a healthy adult in moderate exercise, the amount of nitrogenous food need only be increased by one-tenth.

In connection with the process known as assimilation, by virtue of which food is digested, and part of it is stored up in the body, as a reserve in the shape of fat, or in renewing muscular tissue, it should be noted that there is good reason to believe that this is by no means so simple a process as generally supposed; that the assimilation is not as it were direct, in the sense of the fat in food forming fat in the body, fibrin in the food becoming muscle in the body, and so on, but that complicated processes of the re-arrangement of compounds go on in the animal structure.

The question now arises as to how this oxidation of the food, from which all animal energy is derived, is carried on. It was first pointed out by the German physician Mayer, that "muscle is the instrument by which the transformation of force is accomplished, and not the material which is itself transformed." In fact, according to the most recent researches on this question, all the various chemical changes in food which

we have been considering, appear to be really
carried on in the muscular substance of the body,
through the medium of the blood. The general
facts of the circulation of the blood through
minute vessels in the body, and its maintenance
by the action of the heart, are well known. This
circulation serves a great many purposes, and
the liquid part of the blood carries to the muscular tissues the results of the digestion of the
food in a state prepared for oxidation. The
minute red corpuscles or globules floating in the
blood carry with them the oxygen necessary for
the combustion of this food, and the blood is
distributed over the surface and through the substance of the muscular tissue in very fine vessels.
In some way, which is not yet perfectly understood,
owing chiefly to the difficulties of studying this
question in the living subject, the muscular fibres
absorb both the prepared food and the oxygen,
and in their substance these two re-act upon each
other.

But the function of the blood does not
end here; it carries away the noxious products
of oxidation on its return journey, and takes
them to the lungs, where they are brought in
contact with the fresh air inspired in breathing.
Here an interchange again takes place, the blood
gives off these noxious products, which pass away
in breathing outwards, and it also absorbs a fresh
portion of oxygen, which it again carries to the

interior of the body. There is, therefore, always a stream of oxygen from the blood to the muscles, and of carbonic acid from the muscles to the blood, and the oxygen is *stored up in the tissues.* That this is really the case is proved by the fact that certain animals, *e.g.*, a frog, can produce or give off carbonic acid in an atmosphere free from oxygen.

It appears, therefore, that the oxidation in the body is carried on by the tissues themselves; that the blood is merely a carrier, and the lungs are the vehicle of discharge, and that all this work is necessary for the proper maintenance of life in the higher animals. Hence it has been well remarked that "we live in a world of work, of work from which we cannot possibly escape, and that those of us who do not require to work in order to eat, must at least in some sense perform work in order to live." In connection with the exertion of voluntary muscular power in man, too, it should be noticed that although it acts *under the direction of* the will, and is more or less controlled by it, muscle does not derive its power of acting from the will, any more than a steamboat derives its power of motion from the man at the helm who guides it. The force that propels the boat is derived from the chemical action in the furnace of its steam-boiler (and that again from the sun, as seen in the first part of this chapter), and the motion thus produced is directed by the rudder operated by the steersman, and is controlled by the engineer in charge. So is it with the will, in its relation to the energy of muscular action.

A little attentive consideration of the foregoing pages will show that animals throw off from themselves as the result of the oxidation processes going on in their bodies, precisely those substances which are necessary for the food of plants, viz., carbonic acid, water, and the nitrogenous compound urea, which is speedily changed into a simple salt of ammonia. On the other hand, plants, under the influence of solar energy, decompose these very substances, and from them elaborate food for animals, and whatever is necessary also for their own growth. It is to this wonderful cycle that the term " balance of organic nature " is often given.* A very good example of it

* The following passage from Book II. (Population and Subsistence) of Henry George's " Progress and Poverty," is very suggestive in this connection :—

" For that man cannot exhaust or lessen the powers of nature follows from the indestructibility of matter and the persistence of force. Production and consumption are only relative terms. Speaking absolutely, man neither produces nor consumes. The whole human race, were they to labour to infinity, could not make this rolling sphere one atom heavier or one atom lighter, could not add to or diminish by one iota the sum of the forces whose everlasting circling produces all motion and sustains all life. As the water that we take from the ocean must again return to the ocean, so the food we take from the reservoirs of nature is, from the moment we take it, on its way back to those reservoirs. What we,

is seen in a well-managed aquarium, whether
of fresh or salt water. The animals living there
breathe the oxygen dissolved in the water,
and the resulting carbonic acid also remains in
solution, but is at once seized on by the growing plants that are present, and, as before
stated (p. 175), the liberation of oxygen by
them may often be noticed on a bright day in
the form of a stream of bubbles rising to the
surface; in this way the water may be kept
constantly fresh.

We have now arrived at the conclusion of the
consideration of the special subject which it has
been the object of these pages to set forth—the
mode in which the various forms of Energy are
interchangeable, without any diminution in their
total amount; the manner in which the various
so-called Forces of Nature are related to each
other, or, to speak reverently, how the various

draw from a limited extent of land may temporarily reduce the
productiveness of that land, because the return may be to other
land, or may be divided between that land and other land, or
perhaps all land; but this possibility lessens with increasing area,
and ceases when the whole globe is considered. That the earth
could maintain a thousand billions of people as easily as a thousand
millions is a necessary deduction from the manifest truth that, at
least so far as our agency is concerned, matter is eternal and force
must for ever continue to act. Life does not use up the forces that
maintain life. We come into the material universe bringing
nothing; we take nothing away when we depart. The human
being, physically considered, is but a transient form of matter,
a changing mode of motion. The matter remains and the force
persists. Nothing is lessened, nothing is weakened. And from
this it follows that the limit to the population of the globe can
only be the limit of space."

ways in which the Deity works throughout Nature are mutually interdependent. We have seen reason also to believe that Energy, like Matter, is never destroyed, and that its total amount in the universe is unchangeable, capable of neither increase nor diminution. We know that there are " other systems and other suns, each pouring forth energy like their own, but still without infringement of the law which reveals immutability in the midst of change, which recognises incessant transference or conversion, but neither final gain nor loss. The law generalises the aphorism of Solomon, that 'there is nothing new under the sun,' by teaching us to detect everywhere, under its infinite variety of appearances, the same primeval force. To Nature nothing can be added, from Nature nothing can be taken away; the sum of her energies is constant, and the utmost man can do in the pursuit of physical truth, or in the applications of physical knowledge, is to shift the constituents of the never-varying total. The law of conservation rigidly excludes both creation and annihilation . . . The flux of power is eternally the same. It rolls in music through the ages, and all terrestrial energy, the manifestations of life, as well as the display of phenomena, are but modulations of its rhythm."—(*Tyndall.*)

The first scientific man in this country to draw attention to the doctrine of the correlation of the physical forces was Sir W. (Mr. Justice) Grove, whose work under that title is still one of the best treatises on the subject, but there is no doubt that its truth had been intuitively apprehended by others about the same time, and notably by Dr. Joule and by Dr. Mayer, of Heilbronn, Germany. To enumerate, however, the names of the scientific workers to whose labours the establishment of this doctrine on a firm and unassailable foundation is due, would be an invidious task, and out of place in these pages, which only profess to be a *résumé*, and not a history, of the subject. The treatise on the " Conservation of Energy," by Professor Balfour Stewart, and Faraday's lectures on the various " Forces of Nature" will give fuller information to those who desire it.

Towards the end of his work upon "Correlation and Continuity," Sir W. Grove says :—

" Reviewing the various relations between the various forces we have been considering, it would appear that where one of these is excited or exists, all the others are also set in action ; thus, when a substance such as sulphide of antimony is electrified, at the instant of *electrisation* it becomes *magnetic* in directions at right angles to the lines of electric force ; at the same time it becomes heated to an extent greater or less, according to the intensity of the electric forces. If this intensity becomes exalted to a certain point the sulphide becomes luminous, or *light* is produced; it expands, consequently *motion* is produced, and it is decomposed, therefore *chemical action* is produced . . . This is a strong argument in favour of the theory which regards all these different forces as *modes of motion.*" At another place he says:—"Tracing any force backwards to its antecedents we are merged in an infinity of changing forms of force; at some point we lose it, not because it has in fact been *created* at any one point, but because it resolves itself into so many contributing forces, that the evidence of it is lost to our senses or powers of perception."

Our consideration, then, of the various forms of energy in nature has led us to see—

1. That where one is excited or exists, many others are also set in action; hence, probably, all are modes of motion.

2. That any one can be transformed, either directly or by intermediate steps, into any other.

3. That none of them can be produced but by some other as an anterior.

4. That they act uniformly, *i.e.*, according to fixed laws.

Now the term "Law of Nature," in the scientific sense, is simply the expression of the orderly uniformity with which certain series of phenomena invariably succeed each other, and is not the immediate expression of a personal will. Let us briefly consider the precise meaning that should be attached to the term *law*, used in this connection. To the question, "Why does a stone fall to the ground?" the majority of educated persons would reply, "Because of the law of gravitation," or "Because of the earth's attraction," and would consider that they had given a proper explanation of the phenomenon. A little thought, however, will show that this is clearly no *explanation* at all, but simply an assertion that the stone falls *in accordance with* the law of gravitation; moreover, this law is simply the general expression of the fact that all stones fall towards the earth, or of the still more general fact that all material bodies in the universe tend to fall towards each other; but the law has no coercive power inherent in itself. A good analogy may be drawn from the case of the laws of a state. These are nothing more than the rules laid down by the governing power of that state for the conduct of its members—the expression of their (or its) will; but they have no coercive action in themselves. Mere laws are utterly unable to compel the adoption of a certain line of conduct by any individuals, as is clearly shown in that condition of a country described as "lawless"; the laws exist, but the inhabitants disregard them, and do not regulate their conduct in accordance with them. The expression, "a state governed by law," simply means *governed according to law*, *i.e.*, that the controlling power acts within certain fixed and determinate rules laid down by itself.

Similarly, when we speak of the universe as governed by the laws of a Supreme Ruler, we simply mean that it is governed according to His laws—that these laws are merely the generalised expression of the way in which the Supreme Ruler acts. It is related of the great astronomer Kepler, who discovered the laws of the motion of the planets in their orbits, that when he had perceived their truth, and given this grand generalisation an expression in a form of words, his devout mind gave utterance to an ejaculation of thankfulness at having been permitted " to think the thoughts of God." He felt that he had discovered, as it were, the way in which the Author of the Universe worked—the mode in which he governed.

Hence, so far as pure science is concerned, no law can be anything more than an expression of the fact of the orderly uniformity of the phenomena of the universe; it expresses the relationship of the forces, but gives no clue as to their cause. On the other hand, however, there is nothing in this fact to exclude, as is sometimes maintained, the notion of an intelligent **First Cause.**

The uniformities of Nature, so far from suggesting blind force, have always seemed to me—and I trust that the perusal of these pages may have led others to the same conclusion—to present in their wonderful combination of unity and variety, of harmony and diversity, of grandeur and minuteness, the evidences of a designing mind. As the distinguished man to whom this book is dedicated wrote, more than forty years ago:—"Every step that we take in the progress of generalisation increases our admiration of the beauty of the adaptation, and the harmony of the action, of the laws we discover; and it is in this beauty and this harmony, that the contemplative mind delights to recognise the wisdom and beneficence of the Divine Author of the Universe."

There is, perhaps, no more fitting sentence to close the consideration of this subject, than those noble words in which Sir W. Grove concludes his treatise:—"In all phenomena, the more closely they are investigated, the more are we convinced that, humanly speaking, neither matter nor force can be created or annihilated, and that an essential cause is unattainable. Causation is the will, creation the act, of God."

INDEX.

CASSELL & COMPANY, LIMITED, BELLE SAUVAGE WORKS, LONDON, E.C.

The Magazine of Art. Yearly Volume. With about
400 Illustrations by the first Artists of the day, and beautifully-executed Etching, for Frontispiece. Cloth gilt, gilt edges, 16s.

Some Modern Artists. With highly-finished En-
gravings of their most popular Masterpieces and Portraits. 12s. 6d.

The Forging of the Anchor. With 20 Illustrations
specially executed for the Work, by the first Artists of the day. Small 4to, cloth, gilt edges, 5s.

Picturesque Europe. *Popular Edition.* Vols. I. & II.,
with 13 exquisite Steel Plates, and about 200 Original Engravings in each by the best Artists. Cloth gilt, 18s. each; or the two Vols. in one, 31s. 6d. N.B.—The *Original Edition*, in Five magnificent Volumes, royal 4to size, can still be obtained, price £10 10s.

Egypt: Descriptive, Historical, and Picturesque.
By Prof. G. EBERS. Translated by CLARA BELL, with Notes by SAMUEL BIRCH, LL.D., D.C.L., F.S.A. With Original magnificent Engravings. Cloth bevelled, gilt edges. Vol. I., £2 5s.; Vol. II., £2 12s. 6d.

Picturesque America. Vols. I. & II., with 12 exquisite
Steel Plates and about 200 Original Wood Engravings in each. Royal 4to, £2 2s. each.

Landscape Painting in Oils, A Course of Lessons in.
By A. F. GRACE, Turner Medallist, Royal Academy. With Nine Reproductions in Colour. Extra demy folio, cloth, gilt edges, 42s.

Illustrated British Ballads. With Several Hundred
Original Illustrations by some of the first Artists of the day. Complete in Two Vols. Cloth, 7s. 6d. each; cloth, gilt edges, 10s. 6d. each; morocco, Two Vols., 25s.

Sunlight and Shade. Original and Selected Poems
With exquisite Engravings by the best Artists of the day. 7s. 6d.

Choice Poems by H. W. LONGFELLOW. Illustrated
from Paintings by his Son ERNEST W. LONGFELLOW. Small 4to, cloth, 6s.

The Doré Fine Art Volumes comprise—

	£ s. d.		£ s. d.
Milton's Paradise Lost	. 1 1 0	Don Quixote . .	. 0 15 0
The Doré Gallery .	10 0 0	Munchausen . .	. 0 5 0
The Doré Bible . .	. 4 4 0	Fairy Tales Told Again	. 0 5 0
Dante's Inferno . .	. 2 10 0		

Cassell & Company, Limited: Ludgate Hill, London; Paris; and New York.

Cassell's New Natural History. Edited by Prof.
DUNCAN, M.B., F.R.S., assisted by Eminent Writers. With nearly 2,000 Illustrations. Complete in 6 Vols., 9s. each.

European Butterflies and Moths. By W. F. KIRBY.
With 61 Coloured Plates. Demy 4to, cloth gilt, 35s.

The Book of the Horse. By the late S. SIDNEY. With
Twenty-five Coloured Plates, and 100 Wood Engravings. *New and Revised Edition.* Demy 4to, cloth, 31s. 6d. ; half-morocco, £2 2s.

The Illustrated Book of Poultry. By L. WRIGHT.
With 50 Coloured Plates and numerous Wood Engravings. Demy 4to, cloth, 31s. 6d. ; half-morocco, £2 2s.

The Illustrated Book of Pigeons. By R. FULTON.
Edited by L. WRIGHT. With Fifty Coloured Plates and numerous Engravings. Demy 4to, cloth, 31s. 6d.; half-morocco, £2 2s.

Canaries and Cage-Birds, The Illustrated Book of.
With Fifty-six Coloured Plates and numerous Illustrations. Demy 4to, cloth, 35s. ; half-morocco, £2 5s.

Dairy Farming. By Professor SHELDON, assisted by
eminent Authorities. With Twenty-five Fac-simile Coloured Plates, and numerous Wood Engravings. Cloth, 31s. 6d. ; half-morocco, £2 2s.

Dog, The Illustrated Book of the. By VERO SHAW,
B.A. Cantab. With Twenty-eight Fac-simile Coloured Plates, drawn from Life expressly for the Work, and numerous Wood Engravings. Demy 4to, cloth bevelled, 35s.; half-morocco, 45s.

European Ferns: their Form, Habit, and Culture.
By JAMES BRITTEN, F.L.S. With Thirty Fac-simile Coloured Plates, Painted from Nature by D. BLAIR, F.L.S. Demy 4to, cloth gilt, gilt edges, 21s.

Familiar Wild Birds. By W. SWAYSLAND. FIRST
SERIES. With 40 full-page exquisite Coloured Illustrations and numerous Original Wood Engravings. 12s. 6d.

Familiar Garden Flowers. FIRST, SECOND, and THIRD
SERIES. By SHIRLEY HIBBERD. With Forty Full-page Coloured Plates by F. E. HULME, F.L.S., in each. 12s. 6d. each.

Familiar Wild Flowers. FIRST, SECOND, THIRD, and
FOURTH SERIES. By F. E. HULME, F.L.S., F.S.A. With Forty Coloured Plates and Descriptive Text in each. 12s. 6d. each.

Vignettes from Invisible Life. By JOHN BADCOCK.
With numerous Illustrations specially executed for the Work. Crown 8vo, 3s. 6d.

Cassell & Company, Limited: Ludgate Hill, London; Paris; and New York.

Universal History, Cassell's Illustrated. Vol. I. Early
and Greek History; Vol. II., the Roman Period. With numerous high-class Engravings. Price 9s. each.

England, Cassell's History of. With about 2,000
Illustrations. Nine Vols., cloth, 9s. each ; or in library binding, £4 10s.

United States, Cassell's History of the. With 600
Illustrations and Maps. 1,950 pages, extra crown 4to. Complete in Three Vols., cloth, £1 7s.; or in library binding, £1 10s.

The History of Protestantism. By the Rev. J. A.
WYLIE, LL.D. With 600 Original Illustrations. Three Vols., 4to, cloth, £1 7s. ; or in library binding, £1 10s.

Old and New London. A Narrative of its History,
its People, and its Places. With 1,200 Illustrations. Complete in Six Vols., 9s. each ; or in library binding, £3.

Greater London. Uniform with "Old and New
London." Vol. I. By EDWARD WALFORD. With about 200 Original Illustrations. Extra crown 4to, cloth gilt, 9s.

Our Own Country. An Illustrated Geographical and
Historical Description of the Chief Places of Interest in Great Britain. Complete in 6 Vols., with upwards of 200 Illustrations in each. 7s. 6d. each.

Old and New Edinburgh, Cassell's. Complete in
Three Volumes. With 600 Original Illustrations, specially executed for the Work. Extra crown 4to, cloth, 9s. each.

Krilof and His Fables. By W. R. S. RALSTON, M. A.
Third Edition, Enlarged, 3s. 6d.

The Adventures and Discourses of Captain John
Smith. By JOHN ASHTON. With Fac-similes of the Original Illustrations. Cloth, 5s.

Treasure Island. By R. L. STEVENSON. 304 pages.
Crown 8vo, cloth, 5s.

Technology, Manuals of. Edited by Prof. AYRTON,
F.R.S., and RICHARD WORMELL, D.SC., M.A.
A Prospectus sent post free on application.

Cassell & Company, Limited: Ludgate Hill, London; Paris; and New York.

Cassell's Concise Cyclopædia. A Cyclopædia in
One Volume, containing comprehensive and accurate information, brought down to the Latest Date. Cloth gilt, 15s.

A First Sketch of English Literature. By Professor
HENRY MORLEY. Crown 8vo, 912 pages, cloth, 7s. 6d.

Popular Educator, Cassell's. *New and thoroughly*
Revised Edition. Complete in Six Vols. Price 5s. each.

Science for All. Complete in Five Vols. Edited by
Dr. ROBERT BROWN, M.A., F.L.S., &c., assisted by Eminent Scientific Writers. Each containing about 350 Illustrations. Cloth, 9s. each.

Morocco : its People and Places. By EDMONDO DE
AMICIS. Translated by C. ROLLIN TILTON. With nearly 200 Original Illustrations. Extra crown 4to, *Cheap Edition*, cloth, 7s. 6d.

Energy in Nature. By WM. LANT CARPENTER,
B.A., B.Sc. With 80 Illustrations. 3s. 6d.

The Field Naturalist's Handbook. By the Rev. J. G.
WOOD and THEODORE WOOD. Cloth, 5s.

The Sea : its Stirring Story of Adventure, Peril, and
Heroism. By F. WHYMPER. Complete in Four Vols., each containing 100 Original Illustrations. 4to, 7s. 6d. each. Library binding, Two Vols., 25s.

The Practical Dictionary of Mechanics. Containing
15,000 Drawings, with Comprehensive and TECHNICAL DESCRIPTION of each Subject. Four Vols., cloth, £4 4s.

The Countries of the World. By ROBERT BROWN,
M.A., Ph.D., F.L.S., F.R.G.S. Complete in Six Vols., with about 750 Illustrations, 4to, 7s. 6d. each. Library binding, Three Vols., 37s. 6d.

Peoples of the World. Vols. I. & II. By Dr. ROBERT
BROWN. With numerous Illustrations. Price 7s. 6d. each.

Cities of the World. Vols. I. & II. Illustrated
throughout with fine Illustrations and Portraits. Cloth gilt, 7s. 6d.

Gleanings from Popular Authors. Complete in Two
Vols. With Original Illustrations. Price 9s. each.

Heroes of Britain in Peace and War. By E. HODDER.
With 300 Illustrations. Two Vols., 7s. 6d. each ; or in Library binding, One Vol., 12s. 6d. *Popular Edition*, Vol. I. cloth, 5s.

Cassell & Company, Limited : Ludgate Hill, London ; Paris ; and New York.

The Encyclopædic Dictionary. A New and Original

Work of Reference to all the Words in the English Language, with a Full Account of their Origin, Meaning, Pronunciation, and Use. Five Divisional Volumes now ready, price 10s. 6d. each; or bound in Double Volumes, in half-morocco, 21s. each.

Library of English Literature. Edited by Professor

HENRY MORLEY. With Illustrations taken from Original MSS., &c. Each Vol. complete in itself.
VOL. I. SHORTER ENGLISH POEMS. 12s. 6d.
VOL. II. ILLUSTRATIONS OF ENGLISH RELIGION. 11s. 6d.
VOL. III. ENGLISH PLAYS. 11s. 6d.
VOL. IV. SHORTER WORKS IN ENGLISH PROSE. 11s. 6d.
VOL. V. LONGER WORKS IN ENGLISH VERSE AND PROSE. 11s. 6d.
. Vol. I., *Popular Edition*, now ready, 7s. 6d.

Dictionary of English Literature. Being a Compre-

hensive Guide to English Authors and their Works. By W. DAVENPORT ADAMS. 720 pages, extra fcap. 4to, cloth, 10s. 6d.

Dictionary of Phrase and Fable. Giving the Deriva-

tion, Source, or Origin of 20,000 Words that have a Tale to Tell. By Rev. Dr. BREWER. *Enlarged and Cheaper Edition*, cloth, 3s. 6d.; superior binding, leather back, 4s. 6d.

The Royal Shakspere. A Handsome Fine-Art Edi-

tion of the Poet's Works. Vols. I. & II., containing Exquisite Steel Plates and Wood Engravings. The Text is that of Prof. Delius, and the Work contains Mr. FURNIVALL's Life of Shakspere. Price 15s. each.

Cassell's Illustrated Shakespeare. Edited by CHARLES

and MARY COWDEN CLARKE. With 600 Illustrations by H. C. SELOUS. Three Vols., royal 4to, cloth gilt, £3 3s.

The Leopold Shakspere. The Poet's Works in

Chronological Order, and an Introduction by F. J. FURNIVALL. With about 400 Illustrations. Small 4to, cloth, 6s.; cloth gilt, 7s. 6d.; half-morocco, 10s. 6d.; morocco, or tree calf, 21s.

PRACTICAL GUIDES to PAINTING, with numerous Coloured Plates, and full Instructions by the Artists:—

Flower Painting in Water Colours. FIRST and SECOND SERIES. By F. E. HULME, F.L.S. 5s. each.

Tree Painting in Water Colours. By W. H. J. BOOT. 5s.

China Painting. By MISS FLORENCE LEWIS. 5s.

Water-Colour Painting, A Course of. By R. P. LEITCH. 5s.

Figure-Painting in Water Colours. By BLANCHE MACARTHUR and JENNIE MOORE. 7s. 6d.

Neutral Tint, A Course of Painting in. By R. P. LEITCH. Cloth, 5s.

Sepia Painting, A Course of By R. P. LEITCH. Cloth, 5s.

Cassell & Company, Limited: Ludgate Hill London; Paris; and New York.

The Early Days of Christianity. By the Ven. Arch-

deacon FARRAR, D.D., F.R.S. *Ninth Thousand.* Two Vols., demy 8vo, cloth, 24s. (*Can also be had in morocco binding.*)

The Life of Christ. By the Ven. Archdeacon FARRAR,

D.D., F.R.S.

Bijou Edition, complete in Five Vols., in case, 10s. 6d. the set.
Popular Edition, in One Vol., cloth, 6s.; cloth gilt, gilt edges, 7s. 6d. ;
  Persian morocco, 10s. 6d. ; tree calf, 15s.
Library Edition. 29th Edition. Two Vols., cloth, 24s. ; morocco,
  £2 2s.
Illustrated Edition. With about 300 Illustrations. Extra crown 4to,
  cloth, gilt edges, 21s. ; calf or morocco, £2 2s.

The Life and Work of St. Paul. By the Ven.

Archdeacon FARRAR, D.D., F.R.S. *Library Edition.* (*19th Thou-sand.*) Two Vols., demy 8vo, cloth, 24s. ; morocco, £2 2s. *Illustrated Edition*, cloth, gilt edges, 21s.

An Old Testament Commentary for English Readers.

By various Writers. Edited by the Right Rev. C. J. ELLICOTT, D.D., Lord Bishop of Gloucester and Bristol. Complete in Five Vols., price 21s. each.

  VOL. I. contains GENESIS to NUMBERS.
  VOL. II. contains DEUTERONOMY to SAMUEL II.
  VOL. III. contains KINGS I. to ESTHER.
  VOL. IV. contains JOB to ISAIAH.
  VOL. V. contains JEREMIAH to MALACHI.

A New Testament Commentary for English Readers.

Edited by the Right Rev. C. J. ELLICOTT, D.D., Lord Bishop of Gloucester and Bristol. Three Vols., cloth, £3 3s. ; or in half-morocco, £4 14s. 6d.

  VOL. I. contains the FOUR GOSPELS. £1 1s.
  VOL. II. contains the ACTS to GALATIANS. £1 1s.
  VOL. III. contains the EPHESIANS to the REVELATION. £1 1s.

A Commentary on the Revised Version of the New

Testament for English Readers. By Prebendary HUMPHRY, B.D., Member of the Company of Revisers of the New Testament. 7s. 6d.

Martin Luther the Reformer. By Professor JULIUS

KOESTLIN. With Portraits and other Illustrations. 1s. ; cloth gilt, 2s.

The Bible Educator. Edited by the Very Rev. E. H.

PLUMPTRE, D.D. Illustrated. Four Vols., 6s. each; or Two Vols., 21s.

Cassell & Company, Limited: Ludgate Hill, London ; Paris; and New York.

The Book of Health. A Systematic Treatise for
the Professional and General Reader upon the Science and the Pre-
servation of Health. Edited by MALCOLM MORRIS. With Contribu-
tions by eminent Medical Authorities. Price 21s.

Our Homes and How to make them Healthy. With
numerous practical Illustrations. Edited by SHIRLEY FORSTER
MURPHY, assisted by eminent Contributors. Royal 8vo, cloth, 15s.

The Family Physician. A Modern Manual of
Domestic Medicine. By PHYSICIANS and SURGEONS of the Principal
London Hospitals. Royal 8vo, cloth, 21s.

Medicine, Manuals for Students of.

Elements of Histology. By E. KLEIN, M.D., F.R.S., 6s.

Surgical Pathology. By A. J. PEPPER, M.B., M.S., F.R.C.S.

The Dissector's Manual. By W. BRUCE CLARKE, F.R.C.S., and C. B. LOCKWOOD, F.R.C.S. 6s.

Surgical Applied Anatomy. By FREDERICK TREVES, F.R.C.S. 7s. 6d.

Human Physiology. By HENRY POWER, M.B., F.R.C.S. 7s. 6d.

Clinical Chemistry. By CHAS. H. RALFE, M.D., F.R.C.P. 5s.

Prospectus will be sent post free on application.

The Domestic Dictionary. An Encyclopædia for the
Household. Illustrated throughout. 1,280 pages, royal 8vo. *Cheap
Edition*, price 7s. 6d. ; half-roan, 9s.

Cassell's Dictionary of Cookery. The Largest,
Cheapest, and Best Book of Cookery. With 9,000 Recipes, and
numerous Illustrations. *Cheap Edition*, price 7s. 6d. ; half-roan, 9s.

Choice Dishes at Small Cost. Containing Practical
Directions to Success in Cookery, and Original Recipes for Appetising
and Economical Dishes. By A. G. PAYNE. 3s. 6d.

A Year's Cookery. Giving Dishes for Breakfast,
Luncheon, and Dinner for Every Day in the Year, with Practical In-
structions for their Preparation. By PHYLLIS BROWNE. *Cheap
Edition*, cloth, 3s. 6d.

What Girls Can Do. A Book for Mothers and
Daughters. By PHYLLIS BROWNE, Author of "A Year's Cookery,"
&c. Crown 8vo, cloth. *Cheap Edition*. 3s. 6d.

☞ **Cassell & Company's Complete Catalogue,** *containing a List of
Several Hundred Volumes, including Bibles and Religious Works, Fine-
Art Volumes, Children's Books, Dictionaries, Educational Works, Handbooks
and Guides, History, Natural History, Household and Domestic Treatises,
Science, Serials, Travels, &c. &c., sent post free on application.*

Cassell & Company, Limited; Ludgate Hill, London; Paris; and New York.